D0858234

A LABORATORY MANUAL OF ANALYTICAL METHODS OF PROTEIN CHEMISTRY

VOLUME 5

A LABORATORY MANUAL OF
ANALYTICAL METHODS OF PROTEIN CHEMISTRY

VOLUME 5

EDITORS

P. ALEXANDER
H. P. LUNDGREN

THE QUEEN'S AWARD
TO INDUSTRY 1966

PERGAMON PRESS

OXFORD · LONDON · EDINBURGH · NEW YORK
TORONTO · SYDNEY · PARIS · BRAUNSCHWEIG

Pergamon Press Ltd., Headington Hill Hall, Oxford
4 & 5 Fitzroy Square, London W.1
Pergamon Press (Scotland) Ltd., 2 & 3 Teviot Place, Edinburgh 1
Pergamon Press Inc., Maxwell House, Fairview Park, Elmsford, New York 10523
Pergamon of Canada Ltd., 207 Queens Quay West, Toronto 1
Pergamon Press (Aust.) Pty. Ltd., 19a Boundary Street,
Rushcutters Bay, N.S.W. 2011, Australia
Pergamon Press S.A.R.L., 24 rue des Ecoles, Paris 5^e
Vieweg & Sohn GmbH, Burgplatz 1, Braunschweig

Copyright © 1969
Pergamon Press Ltd

First edition 1969

Library of Congress Catalog Card No. 60–1625

Printed in Great Britain by Thomas Nelson (Printers) Ltd London and Edinburgh

08 012677 4

QD
431
.A63
v.5

CONTENTS

4.66.

95627

PREFACE TO VOLUME 5

THE continued good reception of the previous volumes of the manual encouraged preparation of this added volume to the series. Volume 5 conforms to the general aim and scope of the earlier volumes as stated in the accompanying preface to Volume 4. Volume 5 adds four additional up-to-date laboratory techniques for protein and polypeptide study. In each case the topic is selected on the basis of its wide and growing interest, and is discussed by an eminent authority on the subject.

The Editors wish to thank most sincerely the authors for their cooperation in the preparation of this volume. Also, we are grateful to the editorial staff of Pergamon Press for their continued cooperation.

PREFACE TO VOLUME 4

WHEN the first three volumes were planned there was no intention of continuing the series at periodic intervals. As these manuals have been very well received and appear to fulfil a real need it has been urged upon us that they should be kept up to date. We feel that the best way of achieving this is by the publication at intervals of additional volumes which are not divided into particular types of experimental procedure as were the first three. The contents of these new volumes will be miscellaneous; they will cover topics not dealt with in the first three volumes and will contain new articles on those methods which have undergone significant improvements.

PETER ALEXANDER

September, 1965.   HAROLD P. LUNDGREN

PREFACE TO VOLUMES 1, 2 and 3

IN the last fifteen years there has been a revolution in the techniques available for the analysis and isolation of proteins. Every time a new technique has been introduced, numerous papers have appeared describing modifications to it and the research worker who wishes to employ these methods is faced with the very serious problem in deciding which particular variant to use. These volumes are intended to provide the fullest practical detail so that any scientist can follow the procedure by using this book alone and without having recourse to the original literature. No attempt has been made by the contributing authors to describe all the variants. The techniques which are described in full are ones in which all the authors have had first-hand experience and as a result the descriptions contain those small, but important, points of techniques which are often omitted from the scientific papers, but which save so much time if known. Where the techniques require a large instrument such as the ultracentrifuge or the electron microscope, no attempt has been made to describe the working of these instruments in detail, since this is provided in the manufacturers' manuals. However, the authors have attempted to give full details of the preparation of samples before they can be used in these techniques and for the evaluation of the data. For methods which do not require large instruments or which require instruments which must, in general, be made by the investigator himself, more detailed working details are given. In each of the articles a short discussion of the background and theoretical principle is given and a more detailed description of the difficulties in interpretation. It is our hope that workers who find that they have a problem in protein chemistry will be able to turn to these volumes and, by looking through the chapters, decide which of the techniques is the most suitable for their purpose and then be able to follow this technique from the instructions provided.

In the first volume, separation and isolation procedures are discussed; the second volume concerns its analysis and reactivity, and the third volume with the measurement of the macromolecular properties of proteins.

December, 1960.

R. BLOCK
P. ALEXANDER

LIST OF PREVIOUS CONTENTS

spina-

1

SELECTIVE STAINING OF HISTONES WITH AMMONIACAL-SILVER

By Maurice M. Black, m.d.

Professor of Pathology, New York Medical College

and

Hudson R. Ansley, ph.d.

Assistant Professor of Pathology, New York Medical College

CONTENTS

1

SELECTIVE STAINING OF HISTONES WITH AMMONIACAL-SILVER*

By Maurice M. Black, m.d.

Professor of Pathology, New York Medical College

and

Hudson R. Ansley, ph.d.

Assistant Professor of Pathology, New York Medical College

I. INTRODUCTION

Within the past decade there has been a proliferation of investigations of the basic proteins of cell nuclei, a subject which had been relatively dormant for many years. The renewed interest is evidenced by review papers, symposia and monographs.[1-6] The interest in a class of proteins which were identified and isolated before the turn of the century is due in part to the development of more refined techniques for the analysis of proteins, viz. column chromatography, gel electrophoresis and automatic amino acid analysis. Other technical advances which have proved suitable to the study of nuclear histones are autoradiographic procedures and particularly the demonstration by Alfert and Geschwind that nuclear histones could be stained selectively by acid dyes such as Fast Green (FG) at alkaline pH.[7]

Aside from the technical advances in protein analysis there was also a conceptual interest which arose with the rise of knowledge regarding DNA structure and its genetic function. The demonstration that the sequence of purine and pyrimidine bases of DNA was the biochemical representation of the gene made it pertinent to inquire as to the reason for the ubiquitous association of DNA with histones or protamines in cell nuclei.

It is the purpose of this section to describe a staining procedure for histones which has a high degree of selectivity for basic proteins and the unique ability to visualize qualitative variations in terms of color changes. Accordingly it is a valuable adjunct to the available methods for investigating the nature and function of basic proteins such as the histones and

*Supported in part by Grant CA 05678 from National Cancer Institute, United States Public Health Service.

3

protamines. The procedure in question utilizes ammoniacal-silver (A-S) to stain formalin-treated histones in situ or after isolation.[8–12] The present report will be concerned with a detailed presentation of the procedure, its use and limitations.

Before presenting the details of the procedure, however, it would be appropriate to comment on its development and to distinguish the present technique from the uses of ammoniacal-silver as a histological stain. The reader will probably be aware that ammoniacal-silver has been used for many years as a stain for the cytoplasmic processes of phagocytic cells, viz. the microglia of the central nervous system and the so-called metallophil cells of the lympho-reticuloendothelial system. In fact it was for the latter purpose that one of us (M.M.B.) first used the procedure, following the method of Marshall.[13] Marshall's technique was modified for the staining of routine paraffin sections and applied to studies of lymph node reactivity.[14–19] In the course of such studies we observed that in some instances the nuclei of lymphoid cells rather than the cytoplasm of the RE cells were stained. Such unexpected staining appeared to be related to immunological responses. Furthermore, parenchymal cell nuclei were also stained in special instances whereas cancer cell nuclei stained poorly. These various empirical findings prompted us to investigate the nature of ammoniacal-silver staining, particularly as applied to cell smears. Two fundamental features soon emerged. (1) Staining of cell smears required prior exposure to formaldehyde; no staining occurred without such exposure. (2) After appropriate exposure to formaldehyde, ammoniacal-silver stains the cell nuclei in a highly selective fashion. In contrast, no metallophilic staining of RE cell cytoplasm was demonstrable. Further investigations disclosed that the nuclear component responsible for nuclear stainability with ammoniacal-silver was histone. It thus appeared that under appropriate conditions ammoniacal-silver functions as a cytochemical stain for histones.[8] On the other hand, the metallophilic staining of RE and microglial cell processes appeared to be an artifact of preparation, a reproducible and illustrative technique, but none the less an artifact dependent upon the exposure to ethyl alcohol in the course of the preparation of the slide. Thus cytoplasmic staining is commonly seen in parenchymal cell smears after extraction of nuclear histones with acid and subsequent A-S staining. Some dislodged nuclear histone apparently combines with the acid proteins of the cytoplasm and is stainable with ammoniacal-silver. The cytochemical procedure presented below is to be distinguished from the histological staining procedures wherein ammoniacal-silver is used to demonstrate cytoplasmic metallophilia.

II. A-S STAINING

A. Method

Staining with A-S is influenced by such variables as temperature, pH, duration of fixation, and method of preparing the material, viz. cell smears, cryostat sections, isolated histones. If reasonable attention is paid to these details, the results will be uniform and reproducible.

1. Reagents

(a) *Formalin fixative*
The fixative employed is 10% acetate-neutralized formalin. It is prepared by adding 2 g of sodium acetate to 10 ml of reagent grade formaldehyde (37% formaldehyde) and diluting the solution to 100 ml with distilled water. The fixative is stable and may be kept at room temperature for months.

(b) *Ammoniacal-silver (A-S) solution*
The A-S solution is prepared just before use by adding a 10% aqueous solution of silver nitrate to concentrated ammonium hydroxide until a faint permanent turbidity is obtained. Approximately 40 ml of $AgNO_3$ is required for 3.5 ml of concentrated NH_4OH. It is important to add the $AgNO_3$ solution gradually while stirring with a glass rod.

(c) *Developer*
The reducing solution is 3% formalin prepared by appropriate dilution of the formalin fixative with distilled water.

(d) *Mounting media*
The stained sections and smears are best mounted in a glycerin–gelatin mixture (glycogel). This is prepared by dissolving 15 g of gelatin in 100 ml of warm distilled water, adding 100 g of glycerin and heating for 5 min in a water bath. Filter and use at 37°C. Permount may be used after dehydrating the stained slides in ethanol and clearing in xylol. However, this latter procedure tends to dim the colors and narrow the staining spectrum. Both the glycogel and the permount preparations tend to fade after a few months. Kodachrome II professional film, or Ektachrome film, suitably controlled, give reasonably accurate and permanent records. The colors may be expressed even more precisely in terms of absorption curves measured cytophotometrically (see below).

2. Procedure

(a) *Tissue sections and smears*

Fresh unfixed smears or cryostat sections are air dried at room temperature and then fixed and stained in the following manner, the entire procedure (fixation, staining and developing) being done in a laboratory maintained at 27–29°F. Staining is markedly reduced and erratic, for example, if the temperature of the reagents is lowered to 20°C.

1. The interaction with formaldehyde is obligatory for subsequent A-S staining. The type of staining obtained will vary with the length of formalin fixation. For particular investigations times as short as 1 hr or as long as several days may be most appropriate.
2. Rinse with agitation 7 to 10 times in distilled or deionized water.
3. Immerse with agitation in the A-S solution for 10 sec. The slides may be exposed to the A-S individually or the A-S may be added to the coplin jar containing five slides.
4. Wash with agitation in 7 to 10 changes of distilled water or deionized water.
5. Develop in 3% formalin solution for 2 min.
6. Wash in distilled or deionized water.
7. Mount in glycogel and cover with a coverslip.

The slides are then ready for study. Microscopic examination reveals that cell nuclei are selectively stained by A-S. The color and the intensity of staining varies with the cell type and in some instances with the functional status of the cell. The color range includes yellow, reddish, brown and black (Fig. 1).

A-S stained cryostat sections are usually quite uniform in appearance in all parts of the section. On the other hand, smears are not stained as uniformly. The staining in very thin areas of smears is often very faint while thick clumped areas of the smear may show non-specific staining. The sharpest and most uniform staining is found in that part of the smear where the cells are undistorted and spread evenly in a layer of 1–2 cell thickness. In spite of such staining variables, smears are useful when prepared so that areas of the type described are readily found.

(b) *Isolated histones*

The procedures described above may also be applied to isolated histone fractions. Histones are dissolved in 0.25 N HCl and applied as single drops of the solution to cellulose acetate strips stretched taut between two supports. We have found Oxoid electrophoresis strips (Consolidated Labs.) particularly suitable for these studies. The manufacturer's mark designates the upper surface. A disposable capillary pipette provides a convenient device for obtaining similarly sized drops. The drops are allowed to dry at room temperature before staining.

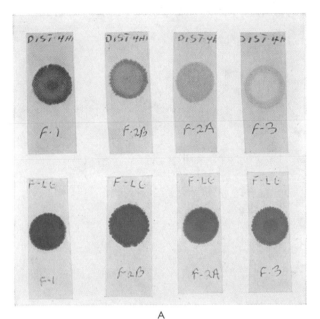

A

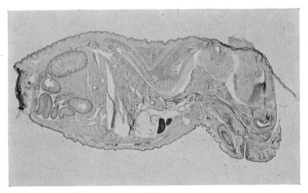

B

Fig. 1. A through H, A-S stained cryostat sections of newborn mouse demonstrate qualitative variations to cell type. A-S stained histone fractions (F-1, F-2b, F-2a and F-3), spotted on paper, demonstrate staining variations in relation to chemical composition. Matching spots stained with Light Green provide a measure of the protein concentration. B, Sagittal section through midline shows A-S colours ranging from black (vertebral column, ribs, intestinal villi) to reddish (thymus), to yellow (lung, kidney).

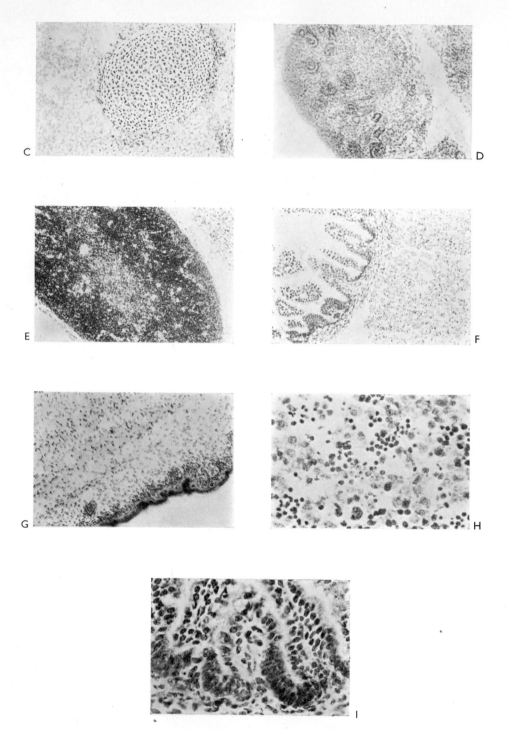

FIG. 1 (*continued*). C, Rib. D, Kidney. E, Thymus. F, Intestine and liver. G, Abdominal wall. H, Liver, showing hematopoietic foci, megakaryocytes and hepatocytes. I, Intestine, showing zones of germinative (yellow) and apical (blackish) cells.

The dried spots are floated on 10% formalin fixative in a Petri dish and the paper allowed to wet from the undersurface up. After wetting is complete the paper is immersed and washed in several changes of the formalin fixative and then left in the fixative overnight. After fixation, the strips are washed with agitation in four changes of distilled water of 15 min each. They are then exposed for 10 sec to freshly prepared A-S solution. After agitation for 10 sec they are then washed vigorously with running distilled water for approximately 30 sec. The papers are then placed in 3% formalin solution for 2 min, rinsed several times in distilled water, blotted and allowed to dry. Appreciable staining of the background paper indicates incomplete washing. Such preparations lack specificity and should be discarded.

The A-S staining of the spots can be shown to be related to the type of histone, viz. Butler's fractions F-1 and F-2b stain yellowish while F-2a and F-3 stain blackish (Fig. 1). The spreading of the drop on the Oxoid paper is influenced by the type of histone in solution. The spreading of the drop into the paper brings about a partial separation of the histones resulting in the formation of rings with varying degrees of prominence. The separation of histone fractions by the spreading is evidenced by differences in A-S staining of the different rings of the same spot (Fig. 2).

(c) *Isolated chromatin fibers*

Chromatin fibers prepared from 2 M NaCl extracts of cell nuclei[20] and affixed to glass slides are stainable with the A-S procedure. The staining of such fibers tends to duplicate that of the nuclei from which they are extracted.

B.   Measurements

1.   Cytophotometry

We have used the Canalco cytophotometer which consists primarily of a digital ratio computer receiving simultaneous impulses from two 1-P-21 photocells, one of which samples light as it emerges from a monochromator while the other measures the same light after it has passed through the object and the optics of a compound microscope. The slits of the monochromator are set to resolve 20 Å of the visible spectrum. Especially well-matched photocells having low dark current and high red sensitivity are utilized. Individual nuclei are measured with appropriately sized plugs which include as much of the nucleus as is possible. Spectral curves are obtained by taking measurements at 20 mμ intervals from 420 mμ to 700 mμ.

Figure 3 depicts some representative curves of A-S stained nuclei

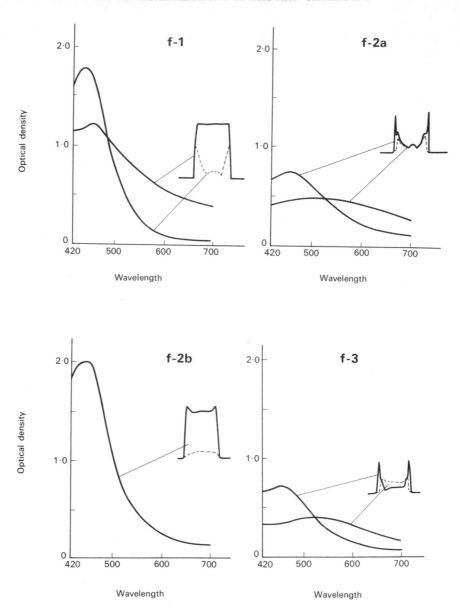

FIG. 2. Spectra of the A-S stained histone extracts spotted on paper with densitometric profiles shown as inserts. In the densitometric inserts the solid line represents the tracings of absorbance at 480 mμ and the dotted line at 610 mμ. As indicated by inserts and spectral lines, F-1, F-2a and F-3 contain differently staining subfractions, whereas F-2b appears more homogeneous. Although the protein concentration of all four spots is quite similar (see Table 1), the intensity of A-S staining varies.

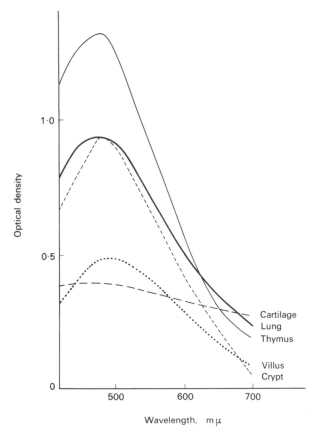

FIG. 3. Spectra of A-S stained nuclei of different cell types of new born mouse, corresponding to type of material illustrated in Fig. 1, show the range of A-S staining.

measured in this way. It is apparent that the curves present characteristic shapes with maxima ranging from 440 to 540 mμ. For example, the yellowish-staining thymocytes show a peak at about 480 mμ and a steep downward slope. Such curves may be characterized more simply by the ration 640/460 mμ.

2. Isolated Histones

Histone extracts, dried and stained on Oxoid paper, may be cleared, mounted on microscopic slides and handled exactly like a histological preparation for microspectrophotometry. We have found mineral oil to be satisfactory as a clearing agent for Oxoid paper. As indicated in Fig. 2, fractions F-1 and F-2b are stained yellowish, whereas fractions F-2a and F-3 are stained blackish; the peak of the A-S curve is shifted to the right. Fraction F-1 is a highly lysine-rich fraction, whereas fraction F-3 is an arginine-rich fraction. Fractions F-2a and b are moderately lysine-rich.

The intensity of staining relative to the concentration of a particular fraction may be roughly determined by comparing the measurements of similar spots stained with A-S and with Light Green. Scanning the Light-Green stained spot at 610 mμ provides a measure of the concentration of the protein within the limits indicated in Fig. 4. Scanning of the A-S

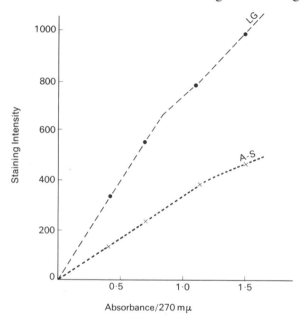

Fig. 4. Relationship between concentration of calf thymus histone as measured spectrophotometrically at 270 mμ and the densitometric scanning of histone spots stained with Light Green (LG) and A-S.

stained spots at absorption maxima provides a measurement of the intensity of A-S staining. As indicated in Table I, the resultant A-S/Light-Green ratios differ for the different fractions, the ratio being greatest for the F-1 fraction and least for the F-3.

C.   Special Procedures

As indicated above, the A-S procedure stains the nuclei of different cell types different colors. This is not to say that all of the different types of nuclei stain a distinctively different color from all other nuclear types, viz. in adult CF-1 mice the nuclei of hepatocytes, pancreatic cells, and

TABLE I. DENSITOMETRIC MEASUREMENTS OF A-S AND
LIGHT GREEN STAINING OF SPOTS OF CALF THYMUS
HISTONE FRACTIONS (WORTHINGTON
BIOCHEMICAL, INC.)*

	Fraction			
	F-1	F-2a	F-2b	F-3
A-S	1135	510	920	320
LG	1035	875	1160	850
AS/LG	1.10	0.58	0.80	0.39

*Prepared as 0.1% solutions in 0.25 N HCl and spotted on
Oxoid cellulose acetate paper.

cerebral neurones all stain blackish whereas nuclei of thymocytes,
cerebellar granular layer cells and breast parenchymal cells stain yellowish. It is possible, however, to further differentiate the similarly stained
cell types, by such procedures as differential extraction, DNA-
ase digestion and blocking reactions. While the possibilities of such
devices have not yet been fully explored some comment is in order at
this time.

1.   Differential Extraction

The use of special solvents to remove particular histone fractions
selectively is a standard biochemical procedure. The variations in the
extractability of nuclear histones of different cell types has been depicted
in previous publications from this laboratory.[6] It is noteworthy that
differential extraction is evident even within different parts of the same
chromosome, viz. polytene chromosomes of dipteran salivary glands.[9]
When squash preparations of such chromosomes are extracted with cold
0.25 N HCl and subsequently stained with A-S it can be seen that histone
is extracted to different degrees in the various bands.

It is common practice in biochemical studies of nuclear histones to
isolate nuclei free of surrounding cytoplasm and then to wash the nuclei
repeatedly to remove non-histone proteins. While such procedures
minimize the contamination of the histone extract they are liable to result
in the loss of certain histone fractions from the nuclei. The reality of
such an effect is readily demonstrable in cryostat sections or cell smears
exposed to repeated washes with 0.9% Na Cl, pH 7.2 and subsequently
stained with A-S. Such preparations show a significant decrease in the
intensity of staining. Furthermore, saline extracts of isolated nuclei
contain acid soluble (0.25 N HCl) proteins which stain like histones with
alkaline Fast Green and A-S and which migrate like histones on poly-
acrylamide gel electrophosesis.

It is apparent from the above observations that extracts which are commonly considered to represent total histone are not necessarily completely representative of the histones in intact nuclei. Physicochemical similarities between histones extracted by similar procedure from different cell types may not necessarily indicate an equivalent similarity of the histones in the intact nuclei. Unless biochemical extraction procedures are monitored by cytochemical observations the aforementioned sources of error will be undetected. Due to its simplicity of performance and its ability to visualize qualitative features the A-S procedure is well suited for such monitoring. At the same time, since histones are easily extracted from some cell types (e.g. thymus) but only with great difficulty from others (e.g. pancreas), such observations provide insight into differences in the binding of histone in different cell types.

2.   Blocking Reactions

Since the A-S staining of histones reflects the interaction of formaldehyde with the ϵ-amino groups of lysine and the guanidino groups of arginine the elimination of such groups should inhibit the A-S stainability. Such effects are readily demonstrable when the cells are treated with nitrous acid (Van Slyke procedure) or with acetic anhydride. Of particular interest, however, is the finding that variations exist in the degree of inhibition of staining of different types of nuclei and within different areas of the same chromosome.[6,9] The procedure employed for the nitrous acid and the acetic anhydride treatment of cryostat sections or smears are as follows:

NITROUS ACID

1. Fix in 10% neutral formalin — 4 hr.
2. Wash in water.
3. Immerse, with agitation, in equal parts of 10% sodium nitrite and 10% trichloroacetic acid. A 5-min immersion will efface much of the yellowish A-S staining but leave varying amounts of blackish staining in different tissues.
4. Wash in water.
5. Wash in 10% formalin and then leave overnight in formalin.
6. Stain with A-S.

ACETIC ANHYDRIDE

1. Fix in 10% neutral formalin — 4 hr.
2. Wash in water.
3. Immerse in acetic anhydride for 2 hr at room temperature.
4. Wash in water.
5. Wash in 10% formalin, then leave overnight in formalin.
6. Stain with A-S.

3. Histone–DNA Interactions

Isolated histone fractions spotted on cellulose acetate paper will combine with DNA and the complex may be stained with A-S or with the Feulgen procedure or both. The procedure which we have employed involves incubating the histone spot in a 0.1% solution of commercially prepared DNA (Nutritional Biochemical Corp.) in 0.9% NaCl, pH 7.2. After a 15-min immersion the histone spot is rinsed with 0.9% NaCl, pH 7.2; three 15-min rinses being used. The spots are then fixed in 10% formalin and stained with A-S. Duplicate spots are stained with the Feulgen procedure.

The DNA will block chromaphilic groups of the histone so that Butler fractions F-1 and F-2b of thymus histone, which ordinarily stain yellowish, stain blackish when complexed with DNA. A-S staining may be altered by DNP complex formation.

One may also stain cells and isolated histone-DNA complexes with the Feulgen procedure after prior A-S staining. Photometric measurements of such doubly stained preparations yields curves which are related to the cell type (Fig. 5). Although there is some influence of the two staining procedures on each other the results are quite reproducible and are therefore of empiric value.

III. COMPARISON OF FG AND A-S STAINING

A. Fast Green Staining

In 1953 Alfert and Geschwind reported that after removal of nucleic acids, the basic proteins of the nucleus could be stained selectively by acid dyes at alkaline pH.[7] At pH 8 the nuclear histones stand out almost, though not quite, alone. The procedure they suggested for tissues is as follows:

1. Fix the tissues for 3–6 hr in 10% neutral formalin.
2. Wash overnight in running water.
3. Dehydrate, imbed and section.
4. Rehydrate sections.
5. Immerse for 15 min in boiling 5% trichloroacetic acid.
6. Wash out the TCA with three changes of 70% ethanol; 10 min each.
7. Wash with distilled water.
8. Stain at room temperature for 30 min in 0.1% aqueous solution of Fast Green (National Aniline Div.) adjusted to pH 8.0–8.1 with minimal amounts of NaOH.
9. Wash sections for 5 min in distilled water.

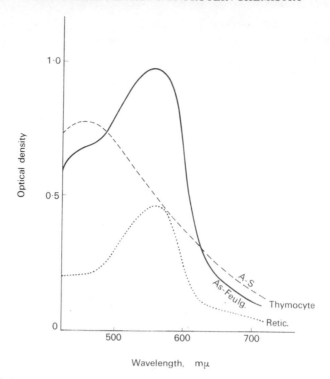

Fɪɢ. 5. Spectra of thymocyte and reticulum cell doubly stained with Feulgen and A-S. Spectrum for A-S staining of thymocyte (same cell) is indicated by thin broken line. The Feulgen procedure reduces the intensity of A-S staining somewhat in the region of peak absorption.

   10. Immerse in 95% ethanol.
   11. Dehydrate, clear and mount.

   The formalin in the procedure functions as a fixative rather than as a reagent since isolated histone spots may be stained directly with FG. Furthermore, the direct extraction of air-dried smears with hot TCA allows for subsequent staining with FG. DNA-ase may be substituted for the hot TCA but RNA-ase is ineffective. The critical feature is the removal of the phosphate acid groups of the DNA so as to make available the previously bound basic groups of the histone.

   Artifactual variations between FG-stained replicate preparations may occur as the result of (a) incomplete removal of DNA, (b) washing out of the FG after staining and (c) change in the pH of the staining solution. In order to provide an internal standard for the staining procedure some investigators include lymphocytes on the test slide, utilizing them as a guide to the adequacy of the stain.

The ability of FG to stain the various histone fractions and combinations thereof in a similar fashion makes it suitable for quantitative cytophotometry. The properties of the dye allow for the use of the two wavelength method.[21] Indirect expressions of qualitative variations may be obtained by the use of preferential blocking or acetylation procedures prior to FG staining.[22,23] More recently, Bloch has reported that eosin may be used to displace FG preferentially from those sites where RNA synthesis is reduced.[24]

The FG staining of nuclear histones *in situ* represents an interaction between the dye and basic groups (ϵ-amino groups of lysine and guanidino groups of arginine) of the histone which were presumably bound to the phosphate acid of the DNA molecule. Although some exceptions have been noted, the resultant staining tends to yield a 1:1 relationship with DNA even in widely different cell types. Thus, the FG method has contributed to the disbelief in histone specificity.

B.   A-S Staining

The A-S procedure does not require the prior removal of DNA but is critically dependent upon the interaction of the histone with formaldehyde. The complexity of reactions between formaldehyde and proteins is well documented in the still pertinent review of French and Edsall.[25] The plethora of possible reactions make it difficult to advance any simple explanation of the reaction between formaldehyde and histones which will completely explain the mechanisms of the A-S staining procedure. Nevertheless, several points appear worthy of emphasis.

(a) There appear to be three types of binding between formaldehyde and proteins, viz. (1) loosely bound formaldehyde which is readily removed by washing in water, (2) chemically bound formaldehyde which resists prolonged water washing and which requires prolonged hydrolysis in hot acid to release the formaldehyde and (3) bound formaldehyde which is removable by prolonged water washing (weeks). It would appear that the latter type of bound formaldehyde would be most likely to interact with the ammoniacal-silver under the conditions employed.

(b) Below pH 8 the binding of formalin to protein is chiefly due to a combination with the ϵ-amino groups of lysine; higher pH being required for reactions with the guanidino groups of arginine. It is noteworthy that Frankfurt has suggested that the A-S staining procedure depends on the disruption by formaldehyde of the bonds between the ϵ-amino groups of lysine and the phosphoric acid groups of DNA. According to his interpretation differences in the A-S staining of different nuclei reflects differences in the avidity of DNA–histone binding.[26]

The yellowish staining of fraction F-1 and the blackish staining of fraction F-3 (see Figs. 1 and 2) would suggest that yellow A-S staining is indicative of a high lysine/arginine ratio whereas blackish staining indicates arginine predominance. However, F-2b, with the lysine/arginine ratio of 1.8, stains distinctly yellow, whereas F-2a, with lysine/arginine ratio of 1.05, stains blackish. These latter observations, while relating yellowish A-S staining with lysine predominance, reveal that the color cannot be explained by any simple stoichiometric relationship.

It is evident from the above that although the A-S procedure involves complex and incompletely understood reactions it does have a basic rationale which is consistent with known physico-chemical interactions between formaldehyde and proteins. Furthermore, the procedure employed would favor staining of lysine histones and be sensitive to subtle variations in the reactivity of the ϵ-amino groups of lysine. Such reactivity could be influenced by interactions with DNA, with other histone fractions and with non-histone proteins or even RNA. The latter possibly is noteworthy in view of a recent report on a labile RNA-histone complex in cell nuclei.[27] Such sensitivity to qualitative variations will perforce limit the application of the A-S procedure as a strictly quantitative index. Accordingly, the FG and A-S procedures are complementary rather than competing procedures.

IV. APPLICATIONS

As mentioned earlier the biochemical literature contains numerous references to the role of histones as gene modifiers or determinants in cellular differentiation. On the other hand, with few exceptions the physico-chemical characterization of histones extracted from different cell types in diverse functional states has revealed an essential similarity which is hardly consistent with a dynamic gene modifying function. Although cytochemical studies using FG or other acid dyes at alkaline pH have demonstrated variations in histones in special instances, FG staining of adult somatic tissue cells is monotonously similar in color.[28,29] In contrast, A-S staining gives evidence of heterogeneity in the chromatin complex of different cell types and more readily discloses changes associated with functional and developmental states. These features are exemplified in the following data:

A.   Antigen-induced Alterations in Lymphoid Cell Histones

The nuclei of normal adult thymocytes and lymph node lymphocytes are characteristically stained yellowish-brown by A-S. However, if an antigen such as tetanus toxoid is injected *in vivo* or administered *in vitro*[10,11,30] an acute reversible change occurs in the A-S staining (Fig. 6). This change consists of a decrease in intensity associated with a transformation from a yellowish to a blackish type of stain. A similar change

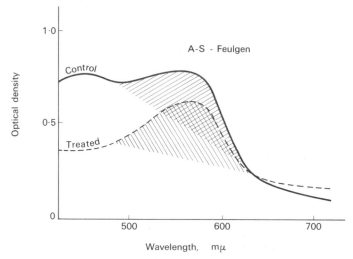

Fig. 6. AS-F spectral curves of thymocytes before and after antigen injection indicate that the A-S color is markedly altered by the antigen treatment without corresponding effect on DNA.

in color is also demonstrable in chromatin fibers prepared from thymocyte nuclei exposed to antigen, either *in vivo* or *in vitro*. However, the effect of antigen on A-S stainability is not demonstrable in isolated histone fraction, e.g. nuclear material extracted in dilute acid such as cold 0.25 N HCl. Furthermore, the antigen-effect is counteracted by prolonged fixation (72 hr) in tepid 10% formalin. The inference is drawn that the antigen-induced alteration in A-S stainability is brought about, not by any quantitative change in histone, but by a shift in histone binding, perhaps to DNA.

An antigen-induced change in the DNP-complex is also suggested by changes in the solubility of isolated chromatin.[31] Chromatin in 2 M NaCl extracts of control thymus glands usually comes out of solution upon dilution to between 0.63 M and 0.57 M. While the test material shows some fiber formation within this range, a significant portion of the DNP is not precipitated until dilutions of 0.54 m to 0.48 M. A-S stains the control fibers a characteristic yellowish color. With the test material, those fibers formed within the control range tend to be stained yellowish brown, whereas those formed only after greater dilution stained blackish.

The above observations demonstrate the ability of A-S staining to visualize subtle changes in the status of nuclear chromatin. The method should prove of particular value on the investigation of the general question of induced alterations in DNA function (induced enzyme synthesis). Thus we have observed changes in the nuclear histones of seminal vesicular cells following castration and testosterone administration.[32]

B. Cellular Differentiation

As indicated above the nuclei of different cell types may be grouped into several categories according to differences in the color of A-S staining. Accordingly, we have studied A-S staining of various organs during intrauterine development of CF-1 mice. The mouse is well suited for such studies since the cytoplasm is unpigmented, yolk is no problem and even the neonate is small enough to be sectioned in total on the cryostat (Fig. 1). Studies on a series of embryos have clearly demonstrated several general points of interest. (a) During the 9th–12th day of gestation the A-S staining is weak and shows no evidence of tissue specific staining. (b) With increasing age all the organs go through a cycle of changes in histone staining which includes the replacement of the faint uniform stain of the early embryo by yellowish staining which increases in

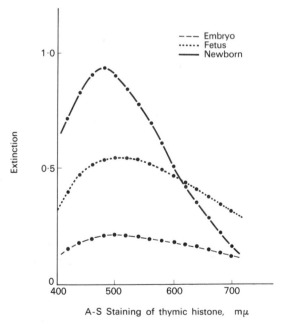

Fig. 7. Spectral curves of mouse thymocytes at various developmental stages demonstrate sequential changes in color and intensity of A-S staining. The change from a faint "undifferentiated" type of stain through a yellowish stage has been noted in various cell types.

intensity with time and is then changed to the characteristic staining color and intensity of the mature cell (Fig. 7). It is noteworthy that these sequential changes do not occur simultaneously in all organs. In some tissues the histone maturation is incomplete for some time after birth.

C. Oncology

The Stedman group suggested that the histones of cancer cells differed in physio-chemical properties from control cell histones.[33] However, no such differences have been found by a number of subsequent investigators using a variety of techniques.[1,2] It has been suggested that apparent differences between control and cancer cell histones reflect differences in the aggregation of histones with non-histone proteins.

In 1960 we reported that cancer cells tend to stain differently from homologous control cells and similarly to each other.[8] In general the cancer cell staining is less distinct than that of the control cells and appears for the most part as diffuse fine granulations with minimal internal differentiation. Frankfurt re-examined this phenomenon and concluded that the differences in A-S staining reflected differences in histone-DNA binding.[26]

It is also noteworthy that there is a definite tendency for the A-S staining of cancer cell nuclei to be less yellow than the homologous control cells (Fig. 8). The possibility that the histone-DNA complex of the cancer cell differs in a subtle fashion from that of the control cell is also suggested by our prior observations on their antigenic properties,[6] and a recent report on the interaction of anti-nuclear antibodies and cancer cell nuclei.[34] While commenting on the findings that suggest that cancer cell histones have particular features in common it should be emphasized that differences in the A-S staining of individual cancers are readily demonstrable. The biological significance of such findings remains to be determined. It appears fair to say that the A-S procedure is a valuable adjunct in the study of the nuclear histones of cancer cells and that the observations thus far indicate a need for further study of cancer cell histones.

D. Mitosis

On the basis of hundreds of observations of A-S stained mitotic nuclei compared with homologous intermitotic cells it may be stated as a general rule that the chromosomes in both states are stained the same color. Only in tumors is there some deviation from this generalization. In our experience the A-S staining of mitotic figures of cancer cells is usually more yellowish than the intermitotic cells (Fig. 8). Whether the application of

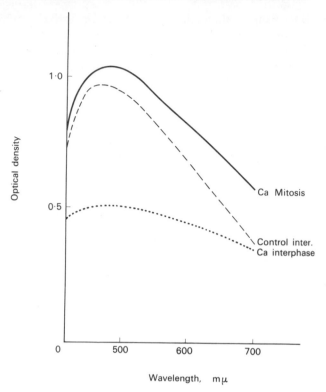

FIG. 8. Spectral curves of A-S stained control and cancerous human breast cell nuclei. The yellowish staining control cell is contrasted with the blackish staining cancer cell. Note that mitotic chromosomes of the cancer cell stain similarly to the control cell nuclei.

special procedures will demonstrate differential features between the chromosomes in different states remains to be determined.

The above applications of the A-S procedure are illustrative. Numerous additional applications will of course suggest themselves.

V. SUMMARY

A formalin-mediated ammoniacal-silver (A-S) staining procedure as a selective stain for basic proteins (histones and protamines) and its application per cell and per extract has been presented. The A-S procedure has been shown to possess unique advantages in investigations of the nature and function of nuclear histones. At the same time, however, its limitations have been indicated and its use as a complementary adjunct to other techniques has been emphasized. When applied to isolated histones the A-S staining varies in a predictable fashion with the type of histone. However, when applied to DNP complexes the staining appears to be related to the relative reactivity of the chromophilic groups (ϵ-amino of lysine and guanidino of arginine). Accordingly the procedure is capable of revealing differences between cell nuclei which are not apparent from studies of isolated histones.

REFERENCES

1. H. Busch and J. R. Davis, Nuclear proteins of tumors and other tissues. Review, *Cancer Res.* **18**, 1241–1256 (1958).
2. J. A. V. Butler, Nuclear proteins of normal and cancer cells. *Exp. Cell Res.* suppl. 9, pp. 349–358 (1963).
3. K. Murray, The basic proteins of cell nuclei. *Ann. Rev. Biochem.* **34**, 209–246 (1965).
4. J. Bonner and P. O. P. T'so, *The Nucleohistones.* Holden-Day Inc., San Francisco, 1964.
5. H. Busch, *Histones and Other Nuclear Proteins*, Academic Press, N. Y., 1965.
6. M. M. Black, H. R. Ansley and R. Mandl, On cell specificity of histones. *Arch. Path.* **78**, 350–368 (1964).
7. M. Alfert and I. I. Geschwind, Selective staining method for basic proteins of cell nuclei. *Proc. Nat. Acad. Sci., U.S.A.* **39**, 991–999 (1953).
8. M. M. Black, F. _ Speer and L. C. Lillick, Acid-extractable nuclear proteins of cancer cell. I. Staining with ammoniacal-silver. *J. Nat. Cancer Inst.* **25**, 967–989 (1960).
9. M. M. Black and H. Ansley, Histone staining with ammoniacal-silver. *Science* **143**, 693–695 (1964).
10. M. M. Black and H. Ansley, Antigen-induced changes in lymphoid cell histones. I. Thymus. *J. Cell Biol.* **26**, 201–208 (1965).
11. M. M. Black and H. Ansley, Antigen-induced changes in lymphoid cell histones. II. Regional lymph nodes. *J. Cell Biol.* **26**, 797–803 (1965).
12. M. M. Black and H. Ansley, Histone specificity revealed by ammoniacal-silver staining. *J. Histochem. Cytochem.* **14**, 177–181 (1966).
13. A. H. E. Marshall, *An Outline of the Cytology and Pathology of the Reticular Tissue.* London, Oliver & Boyd Ltd., 1956.
14. M. M. Black and F. D. Speer, Antigen-induced changes in lymph node metallophilia. *A. M. A. Arch. Path.* **66**, 754–760 (1958).
15. M. M. Black and F. D. Speer, Lymph node structure and metallophilia in tumor bearing mice. *A. M. A. Arch. Path.* **67**, 58–67 (1959).
16. M. M. Black and F. D. Speer, Lymph node reactivity. I. Non-cancer patients. *Blood* **14**, 759–769 (1959).
17. M. M. Black and F. D. Speer, Lymph node reactivity in cancer patients. *Surg. Gyn. & Obst.* **110**, 477–487 (1960).
18. M. M. Black and F. D. Speer, Lymph node reactivity. II. Fetal lymph nodes. *Blood* **14**, 848–855 (1959).
19. M. M. Black and F. D. Speer, Lymph node reactivity. III. Lymphomas and allied diseases. *Blood* **14**, 1026–1032 (1959).
20. C. W. Dingman and M. B. Sporn, Studies on chromatin. I. Isolation and characterization of nuclear complexes of deoxyribonucleic acid, ribonucleic acid and proteins from embryonic and adult tissues of the chicken. *J. Biol. Chem.* **239**, 3483–3492 (1964).
21. L. Ornstein, Distributional errors in microspectrophotometry. *Lab. Invest.* **1**, 250–268 (1952).
22. D. P. Bloch and H. Y. Hew, Changes in nuclear histone during fertilization and early embryonic development in Pulmonate Snail, *Helix Aspersa. J. Biophys. Biochem. Cytol.* **8**, 69–81 (1960).
23. D. P. Bloch and H. Y. Hew, Schedule of spermatogenesis in Pulmonate Snail, *Helix aspersa,* with special reference to histone transition. *J. Biophys. Biochem. Cytol.* **7**, 515–532 (1960).

24. D. P. BLOCH, Differentiation and the regulation of nuclear activity. *J. Cell Biol.* **27**, 12A (1965).
25. D. FRENCH and J. T. EDSALL, The reaction of formaldehyde with amino acids and proteins. *Advances in Protein Chemistry* **2**, 277–335. Eds. M. L. ANSEN and J. T. EDSALL, Academic Press, 1945.
26. O. S. FRANKFURT, The nature of the differences in dyeing nuclei of normal and cancerous cells with ammoniacal-silver. *Vopr. Onkol.* **9**, 61–69 (1963).
27. K. P. CANTOR and J. E. HEARST, Isolation and partial characterization of metaphase chromosomes of a mouse ascites tumor. *Proc. Nat. Acad. Sci.* **55**, 642–649 (1966).
28. M. ALFERT, H. A. BERN and R. H. KAHN, Hormonal influences on nuclear synthesis. IV. Karyometric and microspectrophotometric studies of rat thyroid nuclei in different functional states. *Acta Anat. (Basel)* **23**, 185–205 (1955).
29. M. ALFERT and H. A. BERN, Hormonal influences on nuclear synthesis. II. Volumes deoxyribonucleic acid and protein content of renal nuclei in castrated and testosterone-treated mice. *Rev. Brasil Biol.* **14**, 25–29 (1954).
30. M. M. BLACK and H. ANSLEY, Antigen-induced changes in lymphoid cell histones. III. *In vitro* and in extract. In press. *J. Cell Biol.*
31. M. M. BLACK and H. ANSLEY, Antigen-induced changes in lymphoid cell histones. IV. Changes in solubility of isolated chromatin. In press. *J. Cell Biol.*
32. M. M. BLACK and H. ANSLEY, Induced alterations in nuclear histones. *Fed. Proc.* **24**, 239 (1965).
33. H. J. CRUFT, C. M. MAURITZEN and E. STEDMAN, Abnormal properties of histones from malignant cells. *Nature, London* **174**, 580–585 (1954).
34. T. K. BURNSHAW, T. R. NEBLETT and G. FINE, The immunofluorescent tumor imprint technic. I. Description and evaluation. *Am. J. Clin. Path.* **45**, 714–721 (1966).

2

OPTICAL ROTATORY DISPERSION AND CIRCULAR DICHROISM

By Jen Tsi Yang

from

The Cardiovascular Research Institute and
Department of Biochemistry,
University of California San Francisco Medical Center,
San Francisco, California 94122

CONTENTS

2

OPTICAL ROTATORY DISPERSION AND CIRCULAR DICHROISM

By Jen Tsi Yang

from

The Cardiovascular Research Institute and Department of Biochemistry, University of California San Francisco Medical Center, San Francisco, California 94122

1. INTRODUCTION

Within the past decade optical rotatory dispersion (ORD) has developed into a powerful tool for characterizing the conformations and conformational changes of biopolymers, particularly the proteins. The study of optical rotation began more than 150 years ago, when Biot first observed the rotatory dispersion of quartz, and expanded in the 1840's with the monumental work of Pasteur on the separation of racemic sodium ammonium tartrate. Pasteur's concept of molecular dissymmetry preceded by 25 years the postulation of the tetrahedron model of asymmetric carbon atom. The fertile era of ORD ended with the invention of the Bunsen burner in 1866 which, as Lowry (1935) lamented, made it so easy to work with nearly monochromatic light of the sodium flame that the more laborious study of ORD was nearly abandoned. In the middle 1950's, interest in ORD quickened, partly because good spectropolarimeters became available and partly because of advances in our understanding of the structural elements in the protein molecules. That ORD has a great potential in identifying dissymmetric structural units in a protein or polypeptide molecule is firmly established by the observation of anomalous dispersion of helical polypeptides such as poly-γ-benzyl-L-glutamate and poly-L-glutamic acid (Section 5 (a)). These early investigations prompted the detailed study of the ORD of both polypeptides and proteins. Renewed interest in this method is not so much in the determination of the configurational rotations due to the amino acid residues as it is in the rotatory contributions arising from various conformations. Advances were further stimulated by the appearance of several theoretical treatments of the optical activity of α-helix. It soon became apparent that these early theories were incomplete. In spite of this setback, the Moffitt equation (Section 6 (b)), now regarded as empirical, has been most successfully applied to both polypeptides and proteins.

25

The literature for the 5 years prior to 1961 has been extensively reviewed by Urnes and Doty (1961). Almost all the early work was confined to the wavelength range of 300–700 mμ. Currently, the range of wavelengths has been extended to about 185 mμ, but we have already penetrated some of the absorption bands of the peptide linkages. With the observation of Cotton effects (Section 5 (b)) of various conformations, it is only natural that we are now complementing ORD with measurement of the circular dichroism (CD). The combination of ORD and CD together provides valuable information never before available. Therefore, in this chapter we will describe both ORD and CD, their applications and limitations. Some early quantitative analyses (see Todd, 1960) of the ORD of proteins have been modified or discarded and therefore will only be briefly mentioned.

2. PRINCIPLES AND DEFINITIONS

(a) Optical Rotatory Dispersion (ORD)

A linearly polarized light can be regarded to be made up of two equal, but opposite, circularly polarized components, one left-handed (lcp) and one right-handed (rcp). On passing through an optically inactive medium, the speeds of both components are equally affected, so the plane of polarization remains unchanged upon recombination. But in an optically active medium the speeds of the two components are unequally influenced, so the emergent light rotates through an angle with the plane of polarization of the incident light (Fig. 1). Fresnel in 1825 showed that since the speed of light varies inversely with the refractive index of the medium, optical rotation is simply a phenomenon of circular birefringence (or double refraction). His equation can be expressed as

$$\alpha \,(\text{deg}) = 180 \, l(n_L - n_R)/\lambda, \tag{1}$$

where n_L and n_R are the refractive indices of the left and right circularly polarized light, λ is the wavelength of the incident light, and l is the length of the light path in the medium. The right circular component moves faster than the left one if n_R is smaller than n_L, and vice versa. By convention, rotation is termed positive, or dextrorotatory, if the emergent light is rotated in a clockwise sense as seen by the observer facing the light source and is termed negative, or levorotatory, if the light is rotated counterclockwise. Since the refractive indices are dependent on wavelength, the rotation in Eq. (1) will change with wavelength, thusly optical rotatory dispersion. In general, the magnitude of the rotations increases with decreasing wavelength, provided measurements are made distant

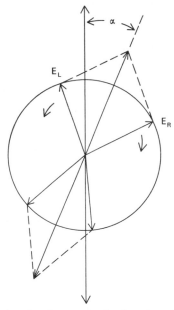

FIG. 1. Resolution of the electric vector of a linearly polarized light into right (E_R) and left (E_L) circularly polarized light. E_R and E_L make unequal angles with the incident light, and the resultant electric vector rotates through an angle of α.

from any optically active absorption bands (see Section 2 (b)). The quantity $(n_L - n_R)$ is exceedingly small compared with the ordinary refractive index, n, which can be taken to be $(n_L + n_R)/2$. For illustration, an α of 1 deg per cm at $\lambda = 540$ mμ would have an $(n_L - n_R)$ of 3×10^{-7}, whereas n is about unity.

Biot in 1836 expressed the optical rotation in terms of specific rotation, $[\alpha]$, and defined it as

$$[\alpha]_\lambda = \alpha/l'C, \tag{2}$$

where α is the degree of rotation at wavelength λ, l' the light path in decimeters, and C the concentration of the optically active substance in gm per ml. Biot used decimeters instead of centimeters in Eq. (2) "in order that the significant figures may not be uselessly preceded by two zeros" (see Lowry, 1935). Thus, the specific rotation calculated from Eq. (2) is ten times what it would be in cgs and its dimension is deg cm^2/decagram. The molar rotation, $[M]$, of a compound is defined, for similar reasons, as

$$[M] = M[\alpha]/100 = M\alpha/100\, l'C \tag{3}$$

(instead of $M[\alpha]$) (M = the molecular weight). The result is $1/10$ what it would be expressed in cgs, and its dimension is deg cm^2/decimole.

Another term, now widely used in polymer chemistry, is the mean residue rotation, $[m]$, which converts the rotation to a monomer residue basis:

$$[m] = M_0[\alpha]/100, \qquad (4)$$

where M_0 is the mean residue molecular weight (for most proteins $M_0 \cong 115$). For quantitative analyses (see Section 6), $[m]$ is further reduced to that under vacuum by a Lorentz correction factor and becomes the reduced mean residue rotation, $[m']$:

$$[m'] = [m][3/(n^2+2)], \qquad (5)$$

where n is the refractive index of the medium. For precise calculations the refractive index dispersion must also be considered, as in the Sellmeir or Maxwell formula:

$$n^2 = 1 + k\lambda^2/(\lambda^2 - \lambda_k^2) \qquad (6a)$$

(k and λ_k are constants characteristic of each solvent used). Equation (6a) can be rearranged to

$$1/(n^2 - 1) = 1/k - \lambda_k^2/k\lambda^2 \qquad (6b)$$

which yields a straight line when $1/(n^2-1)$ is plotted against $1/\lambda^2$ (this formula no longer applies when anomalous refractive index dispersion is observed near the absorption band of the solvent molecule). The numerical values of the Lorentz factor for water and several organic compounds are listed in the Appendix. (N.B. Some workers prefer to use the symbols $[\Phi]$, $[R]$, and $[R']$ instead of $[M]$, $[m]$, and $[m']$ defined in this chapter.)

(b) Circular Dichroism (CD)

Within an absorption band the reduction in intensity of the incident light is characterized by the formula:

$$I = {}_0e^{-4\pi\kappa l/\lambda} \qquad (7a)$$

or, in the more common spectrophotometric practice,

$$I = I_0 10^{-\epsilon m l}, \qquad (7b)$$

where I_0 and I are the intensities of the incident and emergent light at wavelength λ, κ is the absorption index, ϵ is the molar absorptivity (or molar extinction coefficient), m is the concentration in moles per liter, and l is the light path in cm. Since the absorbance (or optical density or absorbancy), A, is defined as $\log_{10}(I_0/I)$, κ is related to ϵ and A by

$$\kappa/\lambda = (2.303/4\pi)\epsilon m \qquad (8a)$$

or

$$\kappa/\lambda = (2.303/4\pi)(A/l). \qquad (8b)$$

If left and right circularly polarized light are absorbed unequally, the sample is said to show circular dichroism. One convenient measure of such CD is the molar quantity, $\epsilon_L - \epsilon_R$, the difference in molar absorptivity for the two circular polarizations.

The presence or absence of CD can also be studied with linearly polarized light. If the left and right circularly polarized components of a linearly polarized beam are absorbed equally, the emergent light remains linearly polarized. If, however, the two components are absorbed unequally, the resultant field no longer oscillates along a single line, and the head of the recombined components traces an ellipse (Fig. 2), that is, the light becomes elliptically polarized. This phenomenon is also called CD. The measurement of CD allows one to classify the absorption bands of

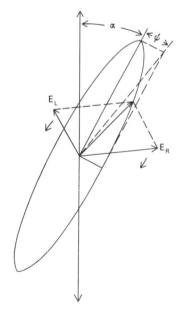

FIG. 2. The rotation, α, and ellipticity, Ψ, of a linearly polarized light through an optically active absorption band. The head of the resultant electric vector traces an ellipse when $E_R \neq E_L$.

a molecule as optically active or optically inactive, depending on whether or not they show nonzero CD.

Since the intensity of light is proportional to the square of the amplitude of its electric vector, E, we have from Eq. (7a)

$$E_L = E_{0L} e^{-2\pi\kappa_L l/\lambda}$$

and

$$E_R = E_{0R} e^{-2\pi\kappa_R l/\lambda} \tag{9}$$

Note that E_{0L} and E_{0R} of the incident light are equal. The ratio of the minor axis $(E_{0R} - E_{0L})$ to the major axis $(E_{0R} + E_{0L})$ of the ellipse in Fig. 2 then defines the tangent of the angle, Ψ, called the ellipticity, i.e.

$$\tan \Psi \text{ (rad)} = (E_R - E_L)/E_R + E_L) \tag{10a}$$

or

$$\tan \Psi \text{ (rad)} = \tanh \pi l(\kappa_L - \kappa_R)/\lambda. \tag{10b}$$

Just as in the case of circular birefringence, the difference $(\kappa_L - \kappa_R)$ is very small compared with the mean absorption index, $\kappa = (\kappa_L + \kappa_R)/2$. Therefore, the ellipse in Fig. 2 is always extremely elongated and we have

$$\Psi \text{ (deg)} = 180 \, l(\kappa_L - \kappa_R)/\lambda \tag{10c}$$

analogous to Eq. (1) for optical rotation. By convention, the ellipticity is termed positive (or negative) if κ_L is greater (or less) than κ_R. Combining Eqs. (8 a, b) and (10 c) and eliminating the κ's, we have

$$\Psi = 2.303 \, (180/4\pi) \, (\epsilon_L - \epsilon_R) \, ml$$
$$= 33 \, (\epsilon_L - \epsilon_R) \, ml \tag{11a}$$

or

$$\Psi = 33 (A_L - A_R). \tag{11b}$$

Just as in the case of optical rotation, we can now define a specific ellipticity, $[\Psi]$ (cf. Eq. (2)), as:

$$[\Psi] = \Psi/l'C \tag{12a}$$

or

$$[\Psi] = 33 (A_L - A_R)/l'C \tag{12b}$$

a molar ellipticity, $[\Theta]$ (cf. Eq. (3)), as:

$$[\Theta] = M[\Psi]/100 \tag{13a}$$

or

$$[\Theta] = 3300 (\epsilon_L - \epsilon_R) \tag{13b}$$

and a mean residue ellipticity, $[\theta]$ (cf. Eq. (4)), as:

$$[\theta] = M_0[\Psi]/100 \tag{14a}$$

or

$$[\theta] = 3300(\epsilon_L - \epsilon_R) \tag{14b}$$

or

$$[\theta] = 3300(A_L - A_R)/ml \tag{14c}$$

and a reduced mean residue ellipticity, $[\theta]$, (cf. Eq. (5)) as:

$$[\theta'] = [\theta][3/(n^2 + 2)]. \tag{15}$$

Here again, l' is in decimeters, l in centimeters, C in gm per ml, and m in moles per liter. The ϵ or A in Eqs. (14 b, c), however, refers to absorbance on the basis of mean residues. The dimension of $[\Psi]$ is again deg cm²/decagram and that of $[\Theta]$ and $[\theta]$ deg cm²/decimole. The expression $(A_L - A_R)$ is directly measurable with the current circular dichrometers, except the Cary Instruments (Section 3(c)) will provide a direct measurement of the ellipticity. Unlike ordinary absorption band, a circular dichroic band can be positive or negative when ϵ_L (or A_L) is greater or less than ϵ_R (or A_R).

(c) Interrelation Between ORD and CD

By utilizing the general Kronig–Kramers relationships, Moscowitz (1960) derived equations for the conversion of CD to ORD and vice versa:

$$[m_i(\lambda)] = (2/\pi) \int_0^\infty [\theta_i(\lambda')][\lambda'/(\lambda^2 - \lambda'^2)] d\lambda' \tag{16}$$

and

$$[\theta_i(\lambda)] = -(2/\pi\lambda) \int_0^\infty [m_i(\lambda')][\lambda'^2/(\lambda^2 - \lambda'^2)] d\lambda'. \tag{17}$$

Computer programs have made solutions of Eqs. (16) and (17) easy. For a partial CD band of the ith electronic transition having a Gaussian form, i.e.

$$[\theta_i] = [\theta_i^0]e^{-(\lambda - \lambda_i^0)^2/(\Delta_i^0)^2}, \tag{18}$$

where θ_i^0 is the extremum of θ_i at wavelength λ_i^0 and Δ_i^0 is the half bandwidth, that is, the wavelength interval over which θ_i falls to $1/e$ (= 0.368) of θ_i^0. By substituting Eq. (18) into Eq. (16), Moscowitz (1960) obtained

$$[m_2] = [2[\theta_i^0]/\sqrt{\pi}][e^{-c^2}\int_0^c e^{x^2}dx - \Delta_i^0/2(\lambda + \lambda_i^0)]. \tag{19}$$

where $c = (\lambda - \lambda_i^0)/\Delta_i^0$. The numerical values of the integral times the exponential in Eq. (19) are listed in the Appendix. Figure 3 illustrates the profile of the first term in the bracket of Eq. (19). This is commonly

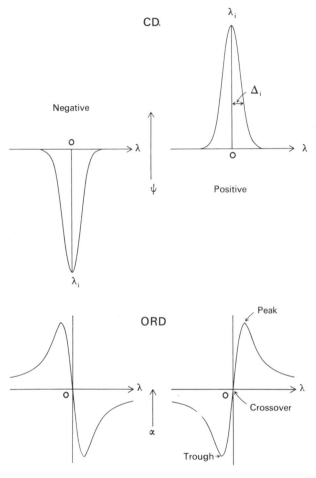

FIG. 3. Idealized Cotton effect of an isolated, optically active absorption band with its maximum at λ_i. The left-half is negative and the right-half positive Cotton effect.

called a Cotton effect in honor of Cotton who in 1895 observed it in optically active absorbing solutions. The Cotton effect is termed positive, if $[\theta_i^0]$ is positive, and the peak (or maximum) of the Cotton effect is on the long wavelength side, and negative when the $[\theta_i^0]$ is negative and the trough (or minimum) is on the long wavelength side. In the U.S.A. the term Cotton effect is reserved for the observation of peaks and troughs in an ORD, but in many European circles it is referred to the phenomena of CD and its associated ORD (Schellman and Schellman, 1964; Velluz *et al.*, 1965).

(d) Rotational Strength

An important theoretical quantity is the rotational strength, R_i, which is defined as the imaginary part of the scalar product of the electric and magnetic transition moment integrals of the ith electronic transition, i.e. $R_i = \mathrm{Im} \, m_i \mu_i \cos \theta_i$ (Kauzmann, 1957). (The intensity of an absorption band depends only on the electric dipole transition moment, whereas the optical activity depends on both the electric and magnetic dipole transition moments. Thus an absorption band may be optically inactive when the said scalar product is zero or active when it is nonzero. Furthermore, a weak absorption band may generate strong optical activity and vice versa.) It is related to the experimental ellipticity by (Moffitt and Moscowitz, 1959; Moscowitz, 1960):

$$R_i = (3hc/8\pi^3 N_1) \int_0^\infty [\Psi_i(\lambda)/l\lambda] \, d\lambda. \tag{20a}$$

Here h is the Planck constant, c the speed of light in vacuum, and N_1 the number of optically active molecules per ml. The coefficient $(3hc/8\pi^3)$ equals 2.40×10^{-18}, the ellipticity per unit length is in radians per cm, and the dimension of R_i is erg cm³ rad. Equation (20a) can also be expressed in terms of the mean residue ellipticity as

$$R_i = 0.696 \times 10^{-42} \int_0^\infty [\theta_i(\lambda)/\lambda] \, d\lambda \tag{20b}$$

noting that $\Psi/lN_1 = \pi[\theta]/18N$ or 2.90×10^{-25} (N = Avogadro's number; cf. Eqs. (12a), (14a)).

For a Gaussian CD band, Eq. (20b) reduces to

$$R_i \cong 0.696 \times 10^{-42} \sqrt{\pi} [\theta_i^0] \Delta_i^0 / \lambda_i^0$$

$$\cong 1.234 \times 10^{-42} [\theta_i^0] \Delta_i^0 / \lambda_i^0. \tag{21}$$

By substituting Eq. (21) into Eq. (19), we have

$$[m_i] \cong (R_i/0.696 \times 10^{-42})(\lambda_i^0/\Delta_i^0)(2/\pi)[e^{-c^2} \int_0^e e^{x^2} dx - \Delta_i^0/2(\lambda + \lambda_i^0)]$$

$$\cong (0.915 \times 10^{42} R_i \lambda_i^0 / \Delta_i^0)[e^{-c^2} \int_0^e e^{x^2} dx - \Delta_i^0/2(\lambda + \lambda_i^0)]. \tag{22}$$

Thus, for a given set of λ_i^0 and Δ_i^0, it is the rotational strength, R_i, that determines both the sign and magnitude of the partial mean residue (or molar) rotations.

As $(\lambda - \lambda_i^0)/\Delta_i^0$ goes to $\pm\infty$, the above integral times the exponential approaches asymptotically $\Delta_i^0/2(\lambda - \lambda_i^0)$. Equation (22) in turn is approximated by

$$[m_i] \cong (0.915 \times 10^{42} R_i)(\lambda_i^0)^2/(\lambda^2 - (\lambda_i^0)^2). \qquad (23a)$$

Thus, in regions distant from the ith CD band, the corresponding ORD simply reduces to a Drude equation (see Section 6 (a)), which can be rewritten as

$$[m_i] = a_i \lambda_i^2/(\lambda - \lambda_i^2). \qquad (23b)$$

(For brevity, the superscript ^{0} is omitted in Eq. (23b).) The magnitude of the rotations associated with any CD band falls off rather slowly on both sides of the band center according to Eq. (23b), unlike the CD curve, which drops exponentially for a Gaussian form (Fig. 3). Therefore, even in the region of the Cotton effect the rotatory contributions from CD bands other than the one under consideration (the so-called background rotation) may be appreciable. Ignoring the background rotation can sometimes lead to erroneous interpretation of the ORD results.

Carver *et al.* (1966a) have used an asymptotic expansion to suggest that the integral times the exponential in Eq. (22) can be better approximated as

$$e^{-c^2} \int_0^c e^{x^2}\, dx \cong \Sigma (4/c)^{2n-1} a_{2n-1}, \qquad (24)$$

the first three coefficients being $a_1 = 0.12499$, $a_3 = 0.00392$, and $a_5 = 0.00031$. Instead of Eq. (22), they introduced an error term to the Moscowitz equation, that is,

$$[m_i] \cong (0.915 \times 10^{42} R_i)[(\lambda_i^0)^2/(\lambda^2 - (\lambda_i^0)^2) + \lambda_i(\Delta_i)^2/4(\lambda^2 - (\lambda_i)^2)^3]. \qquad (25)$$

This correction is very small.

3. INSTRUMENTATION

(Section 3 is by GEORGE M. HOLZWARTH)

A critical evaluation of CD and ORD curves requires knowledge of the precision and accuracy of the curves and thus also of the optical and electronic principles used in the instrument generating the data. To facilitate such evaluation, three topics are examined in the present section:

1. The limitations of precision and accuracy imposed upon CD and ORD curves by the instrument.
2. The optical and electronic principles used in several commercially available spectropolarimeters.

3. The optical and electronic principles used in several instruments, termed "circular dichrometers", for the measurement of CD.

(a) Precision and Accuracy

The precision of the CD or ORD indication is determined by an accumulation of noise from several sources. The dominant source, especially in the ultraviolet regions, is frequently "shot noise" determined by the statistical distribution of photoelectrons generated in the photomultiplier by the light beam (Fried, 1965; Grau, 1965). This noise will vary as the square root of the number of photons received by the phototube during a measurement period. Consequently, by using wide slit widths, low sample absorbances, long integration times, and light sources of maximal intrinsic brightness, this noise source can be partially controlled. A second common source of noise arises from variations in the position or intensity of the light source. The operator of an instrument has little control over this noise source except to insure that a stable arc is properly aligned in the instrument. Other noise sources include Johnson noise in the first stages of electronic amplification, and noise originating in mechanical vibrations of optical elements. In a well-designed instrument noise from these sources should be negligible compared to shot noise.

Like precision, the accuracy of the CD and ORD records reflects both instrument design and proper instrument use. Simple tests for accuracy in CD instruments are not available, but linearity and freedom from stray light can be checked by application of the Beer–Lambert law (Gillam and Stern, 1957). ORD instruments on the other hand, are readily checked for accuracy with standard sugar solutions as well as through the Beer–Lambert-type test. Interconnected with CD or ORD accuracy is the need for adequate spectral resolution, a requirement identical to that encountered in absorption spectrophotometry. However, the broader slit widths commonly employed for reasons of precision in measurements of optical activity make more probable the distortion of ORD and CD curves by inadequate spectral resolution. This is readily checked by repeating a measurement at several spectral bandwidths.

Unlike precision and accuracy of the CD or ORD indication, the precision of the wavelength indication of a spectropolarimeter or circular dichrometer record is usually not under control of the operator and is, for solution work, rarely critical. Accuracy, on the other hand, may be reduced by excessive scanning rates. The instrument indication itself can be checked by inserting in the beam an absorber with sharp absorption lines occurring at known frequency and noting the wavelength reading at which the resultant alterations in light intensity occur.

(b) Spectropolarimeters

The function of the spectropolarimeter is to provide an accurate and precise measurement of the optical rotatory power of the sample over the broadest possible spectral range within as short a time as possible and in the presence of substantial light absorption by the sample. Although different spectropolarimeters carry out this function in different ways, certain fundamental optical and electronic principles may usefully be considered together here.

1. The Determination of the Direction of Vibration of Linearly Polarized Light

The determination of the plane of vibration of a linearly polarized light beam can be achieved by a variety of methods (Lowry, 1935; Cary et al., 1964), but we shall here consider only one, the method of symmetrical modulation, which lends itself to automatic, high-precision measurements.

In this method, the polarizer is caused to oscillate sinusoidally through a few degrees $\Delta\Phi$ about a mean angle α, extinction occurring for α equal to zero. Thus Φ equals $(\Delta\Phi \sin 2\pi ft + \alpha)$, and the transmitted intensity I is given, from Malus' law, that I equals $I_0 \sin^2 \Phi$, by

$$I = I_0 \sin^2 (\Delta\Phi \sin 2\pi ft + \alpha). \tag{25a}$$

If now $\Delta\Phi$ and α are both small angles, then, using the relation $\sin \Phi \approx \Phi$, one has

$$I = I_0(4\alpha\Delta\Phi \sin 2\pi ft - \Delta\Phi^2 \cos 4\pi ft + \Delta\Phi^2 + 2\alpha^2)/2. \tag{25b}$$

There are two terms of interest in this expression. The first term, $2\alpha\Delta\Phi \sin 2\pi ft$, is *linear* in the error α and is modulated at the frequency f. This term will disappear if the polarizer oscillates about the angle defined by $\alpha = 0$. The second term does not depend upon α and is generally considerably larger than the first. However, it can easily be separated from the first because it is modulated at frequency $2f$. The signal at frequency $2f$ can be used as a convenient monitor of the light intensity I_0. The third and fourth terms are time-independent and can thus be eliminated, if this is desired.

The value of the method of symmetrical modulation lies not in its solution to the optical problem alone, but rather in the electrical and mechanical techniques which can be married to the optical system so as to adjust automatically the polarizer angle or the direction of polarization of the incoming beam to achieve a precise null. At the heart of this electrical technique is the method of phase-sensitive detection (Wilson, 1952), which is ideally suited to the isolation of the signal $2\alpha\Delta\Phi \sin 2\pi ft$ from noise. Phase-sensitive detection is in this instance equivalent to multiplication of the alternating electrical signal by $\sin 2\pi ft$. If a time average of the detected signal, plus noise, is obtained, the average of $2\alpha\Delta\Phi \sin^2 2\pi ft$ is simply $\alpha\Delta\Phi$. In contrast to the signal at frequency f, however, noise at frequencies other than f and its odd harmonics will tend to average to zero. Consequently, the averaged output from the phase-sensitive detector, which is largely free of noise and linearly dependent on α, can be used to control a servo-motor or amplifier which adjusts the optical system so as to minimize the signal arising from nonzero α. The optical and electronic systems thus form a closed, self-adjusting loop; this is precisely what is required for an automatic, self-nulling polarimeter. All commercial recording spectropolarimeters utilize this technique.

2. The Faraday Cell

The Faraday cell is a device which can be used to rotate the plane of polarization of a linearly polarized light beam so as to compensate for sample rotation or provide symmetrical modulation of the polarization angle.

Normally, when a beam of linearly polarized light is passed through a transparent slab of isotropic material whose end faces are perpendicular to the direction of propagation of the beam, the beam is unaltered in its properties except for slight scattering and reflection losses. If a magnetic field is now introduced in such a way that the field is parallel to the propagation vector of the beam, the direction of vibration of the linearly polarized beam is altered but the beam properties are otherwise unaffected. This property of isotropic matter is termed the Faraday effect and the rotation induced, θ, obeys the relation, $\theta = VHL$. Here V is the Verdet constant, which is dependent upon the material used and upon the wavelength of the light; H is the magnetic field strength in the direction of propagation of the beam; and L is the path length of the beam in the slab. The magnetic field may be generated by a current-carrying coil to yield a device, termed a Faraday cell, which produces a rotation which varies linearly with a current. A time-independent current can be used to compensate the rotation introduced into the system by the sample; an alternating current can provide the symmetrical excursions of the direction of vibration of the polarized beam required by the method of symmetrical modulation discussed immediately above.

3. Commercial Spectropolarimeters
(i) *Bellingham and Stanley, Ltd., and Bendix Electronics Ltd.*

A novel spectropolarimeter, based on the design of Gillham and King (1961), is manufactured in England jointly by Bellingham and Stanley, and Bendix Electronics, as the Polarmatic 62. The polarimeter and mono-chromator are designed as a single unit utilizing crystal quartz prisms both to disperse the radiation and to polarize or analyze it. A greatly simplified optical diagram of the instrument is shown in Fig. 4. The instrument operates as follows. Light from the arc A passes through slit S_1 and is reflected from mirror M_1 to prism P_1. Prism P_1 acts as both linear polarizer and monochromator; the wavelength setting is fixed by the angular position of mirror M_1. The beam, consisting of two ortho-gonally polarized rays characterized by wavelengths λ_1 and λ_2 determined by the ordinary and extraordinary refractive indices of the crystal

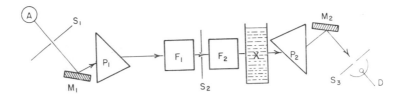

Fig. 4. Optical block diagram of the Polarmatic 62 polarimeter. The light source is designated A; S_1, S_2, and S_3 are slits; M_1 and M_2 are mirrors; P_1 and P_2 are crystal quartz prisms to disperse and polarize the radiation; F_1 is a compen-sating Faraday cell; F_2 is a polarization modulating Faraday cell; X is the sample; and D is the detector.

quartz, then passes through Faraday cells F_1 and F_2, which compensate for sample rotation and modulate the direction of vibration of the light at 380 cps, respectively. An intermediate slit S_2 is placed between the two Faraday cells to improve the spectral purity of the monochromator. After being modulated at F_2, the beam traverses the sample X, where it may be rotated and partly absorbed. The attenuated and rotated beam emergent from the sample then passes via prism P_2 and mirror M_2 to exit slit S_3. The prism disperses the radiation and, in addition, acts as an ana-lyzer of the polarized radiation falling upon it. Mirror M_2 is set so that only the beam at λ_1 is transmitted by S_3. Finally, the beam signal is transduced into an electrical signal by the photomultiplier D.

The electronic portions of the instrument are simple in concept. The current from the photomultiplier contains signals at 380 and 760 Hz(cps), as well as noise. The 380 Hz signal is amplified, filtered, and passed through a phase-sensitive detector. The resultant signal is used to control a d.c. amplifier providing a compensating current to Faraday cell F_1, so as to nullify the 380 Hz signal. The current required by the compensating cell is a direct measure of the optical rotation and can thus be recorded. For a given rotation this current is dependent also upon the wavelength, because the Verdet constant of F_1 shows dispersion. An automatic correction for this factor is incorporated in the instrument.

There are several obvious advantages to this design, notably its use of relatively few optical elements and even fewer moving parts. The wavelength range is 181–600 $m\mu$.

(ii) *Cary Instruments*

Cary Instruments manufactures a recording spectropolarimeter, designated Cary Model 60, whose features have been described in detail in the literature (Cary *et al.*, 1964). A simplified optical block diagram of the instrument is shown in Fig. 5. The instrument operates as follows.

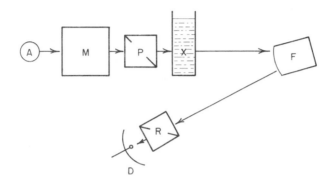

FIG. 5. Optical block diagram of the Cary 60 polarimeter. The symbol A designates the light source; M is the monochromator; P and R are the polarizer and analyzer; X is the sample; F is a Faraday polarization modulator; D is the detector.

Light from the arc A is rendered monochromatic by a fused-silica double monochromator M and linearly polarized by the Senarmont polarizer P. The beam then passes through the sample cell X and enters a silica Faraday cell F driven at 60 Hz. The beam is reflected from the back of the Faraday cell and emerges with twice the modulation angle which the silica would give upon a single passage. The beam then travels via a fixed Rochon analyzer R to the photomultiplier tube. The spectral range is 185 to 600 $m\mu$.

The electrical signals are manipulated as follows. The signal is first amplified and the large unwanted 120 Hz signal removed by filters. The 60 Hz signal, which indicates the error in the null position of the light beam falling on the analyzer, is amplified, passed through a phase-sensitive detector, and finally used to control a servo-motor which rotates the polarizer through a special linkage to achieve a null. The angular position of the polarizer controls a 40-turn potentiometer which is used to indicate the polarizer angle to the recorder.

(iii) *Japan Spectroscopic Company*

A simplified optical diagram of the Jasco ORD-5 instrument is shown in Fig. 6. The instrument functions as follows. The arc A provides radiation to the fused-silica-prism double monochromator M. The monochromatic emergent beam is linearly polarized by the quartz Rochon P_1

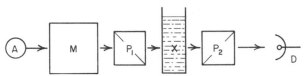

FIG. 6. Optical block diagram of the Jasco, Perkin-Elmer, and Rudolph polarimeters. The symbol A designates the light source; M is the monochromator; P_1 and P_2 are polarizer and analyzer; X denotes the sample; and D is the detector.

which oscillates about its axis through 2° at 12 Hz. The modulated beam then passes through the sample X to the analyzing prism P_2, also a quartz Rochon, and thence to the photomultiplier tube D.

The electronic signal from the photomultiplier contains 12- and 24-Hz signals; the 12-Hz error signal is isolated by phase-sensitive detection and used to control the angular setting of the analyzer through a servo-motor, worm gear, and inclined plane system. Finally, the rotation is recorded as the position of the inclined plane along the direction perpendicular to the analyzer axis.

The principles used in the Jasco generate considerable mechanical complexity in comparison to the Cary 60 and the Polarmatic 62. However, the use of mechanical modulation and compensation eliminates the need for Faraday cells. The spectral range for ORD measurements is 185–700 mμ.

(iv) *Perkin-Elmer*

The Perkin-Elmer Model P 22 spectropolarimeter also possesses an optical block diagram which is described by Fig. 6. In this instrument, light from the arc A is passed through a double-grating monochromator M and then falls upon a calcite Glan polarizer P_1 which oscillates about its axis at 60 Hz through a small angle. The light beam then traverses the sample X and falls upon the calcite analyzer P_2. Finally, the light transmitted by P_2 is detected by the photomultiplier D. The electronic signals from the detector are processed to drive a servo-motor which in turn rotates the analyzer to achieve an optical null, thus completing the signal loop to which the method of symmetrical modulation lends itself.

The use of gratings in this instrument permits mechanical simplifications both in the wavelength presentation and in the slit system, which is wavelength-invariant. However, the use of calcite polarizers restricts the wavelength range to 210–600 mμ.

(v) *Rudolph Instruments Engineering Company*

The Rudolph recording spectropolarimeter was probably the first commercially available instrument capable of measurements into the far ultraviolet region. Its simplified optical diagram is identical to that of the Jasco ORD-5 shown in Fig. 6. The precision of this instrument is considerably lower than that of the three instruments described above, but it is capable of measuring very large rotations, up to 200°, without linearity errors.

(vi) *Zeiss*

Information on the Zeiss instrument was limited at the time of this writing, as the instrument was not yet available in the United States. The few details available are as follows.

The instrument uses a 450-W xenon lamp and the standard Zeiss MM 12 double monochromator as a source of monochromatic light. The polarizer and analyzer are of the Glan type; a silica Faraday modulator is employed. The automatic compensation method is not known, but the design appears to be similar to that of the Cary 60 in optical layout, although presumably without the folded Faraday cell of the latter. Over the usable spectral range 215–650 mμ, an accuracy of 1 millidegree is claimed.

(c) Circular Dichrometers

Circular dichrosism has been measured by two distinct techniques, termed "ellipsometry" and "direct absorbance difference spectrophotometry". Ellipsometry involves measurement of the extent to which the sample alters a linearly polarized light beam into a slightly elliptically polarized beam (see Section 2 (b) above). This method is ill suited to automatic measurements over a broad spectral range and no commercial high-precision instruments utilizes this approach. The reader interested in ellipsometry is referred to Lowry (1935). In the direct absorbance difference method, left circularly polarized light (lcp) and right circularly polarized light (rcp), separated from one another either in space or in time, are passed through the sample and the absorbance difference is recorded. Most of the optical elements utilized in circular dichrometers, such as light sources, monochromators, polarizing prisms, and photomultiplier detectors, are described in standard texts on optics. However, two supplementary elements, namely phase plates and circular polarizers, are less familiar.

1. The Phase Plate

The phase plate is generally a thin slab of birefringent material cut from a crystal with an optic axis oriented in the plane of the plate. The plate therefore introduces phase shifts between light beams of different linear polarization direction when these pass through the plate. The operation of the phase plate is shown diagrammatically in Fig. 7 in which a co-ordinate system $\hat{s}, \hat{f}, \hat{z}$ is set up with optic axis along $\hat{f}$. For light with

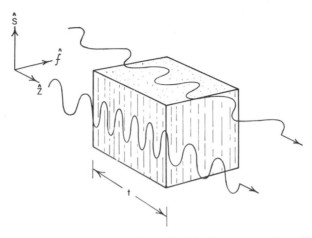

FIG. 7. Schematic diagram of the effect of a birefringent crystal on the relative phase of two orthogonally polarized beams of identical frequency. The beam with vibrations along $\hat{s}$ encounters a higher refractive index than the beam with vibrations along $\hat{f}$. Consequently, if the phases of the two beams are identical where those of the other beam upon emergence from the crystal.

vibration direction along $\hat{f}$ and propagating in the $\hat{z}$ direction, the optical thickness of the plate, T_f, i.e. the vacuum distance spanned by the same number of waves required to span a plate of thickness t, is given by the relation $T_f = tn_f$, where n_f is the refractive index for light with vibration direction along f. For light with vibration direction along $\hat{s}$, similarly, the optical thickness T_s will be equal to tn_s. If n_f is unequal to n_s, the plate is birefringent, and $(T_f - T_s)$ will equal $t(n_f - n_s)$. Consequently, two waves, initially in phase and incident along $\hat{z}$ but with vibration directions along $\hat{f}$ and $\hat{s}$ respectively, will differ in phase upon emerging from the plate by $2\pi(T_f - T_s)/\lambda$ radians. If the quantity $2\pi(T_f - T_s)/\lambda$ equals $\pi/2$, the phase plate is termed a "quarter-wave plate"; $2\pi(T_f - T_s)/\lambda$ equal to one-half characterizes the half-wave plate, etc. This phase difference is shown in the figure.

The birefringence $(n_f - n_s)$, which is the physical property on which the phase plate depends, may be natural, as observed in crystalline quartz, calcite, and mica, or it may be induced by an electric field, as in KH_2PO_4 and similar materials, in which case the phase plate is termed an electro-optic light modulator (EOLM) or Pockel cell (Velluz, et al., 1965; Billings,

1949). For the measurement of circular dichroism, the EOLM possesses advantages over other types of phase plates because the magnitude and sign of the retardation can be varied in time by alternating the electric field.

2. The Circular Polarizer

The present section contains a description of the manner in which a birefringence phase plate can be combined with a linear polarizer to yield a circular polarizer. Just as linearly polarized light may be described as a sum of circularly polarized beams, so circularly polarized light can be depicted as a sum of linearly polarized beams. Thus, circularly polarized light propagating in a direction $\hat{z}$ can be described as a sum of two coherent beams exhibiting orthogonal linear polarization along the x and y directions, as follows:

lcp: $E = E_0 \cos (\omega t + \Phi + 2\pi z/\lambda)(\hat{x} + E_0 \cos (\omega t + \Phi + 2\pi z/\lambda - \pi/2)\hat{y},$

$$(26a)$$

rcp: $E = E_0 \cos (\omega t + \Phi + 2\pi z/\lambda)\hat{x} + E_0 \cos (\omega t + \Phi + 2\pi z/\lambda + \pi2) \hat{y}.$

$$(26b)$$

Here $\hat{x}$, $\hat{y}$, and $\hat{z}$ form a Cartesian coordinate system, and the two terms in each expression describe light linearly polarized along $\hat{x}$ and $\hat{y}$. This formulation of circularly polarized light in terms of two coherent orthogonal linearly polarized beams just 90°, or one-quarter of a wave, out of phase, provides the key to understanding the operation of the Fresnel circular polarizer consisting of a linear polarizer and quarter-wave plate in series, which is shown diagramatically in Fig. 8. Unpolarized but parallel light from the source travels down the $\hat{z}$-axis to the linear polarizer.

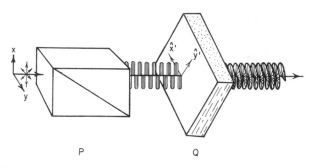

Fig. 8. Schematic diagrams of a circular polarizer. Unpolarized light emanating at the left from the coordinate origin is rendered linearly polarized by the prism P. The plane polarized beam then traverses the quarter-wave plate Q and emerges as circularly polarized light at the right.

The light emergent from the linear polarizer may be written in the form

$$E = E_0 \cos (\omega t + 2\pi z/\lambda + \Phi)\hat{x}. \tag{27}$$

The phase plate is oriented so that its optic axis, y', makes an angle of 45° with the x-axis. The components of the beam incident on the phase plate along the two directions x' and y' are thus of equal intensity and identical phase. If the plate is a quarter-wave plate with a phase difference of $\pm$ one-quarter wave, then the electric fields $E'_{x'}$ and $E'_{y'}$ of the light emergent from the plate will be, neglecting reflection effects,

$$E'_{x'} = E_0 \cos (\omega t + \Phi' + 2\pi z/\lambda)\hat{x}'/\sqrt{2}, \tag{28a}$$

$$E'_{y'} = E_0 \cos (\omega t + \Phi' + 2\pi z/\lambda \pm \pi/2)\hat{y}'/\sqrt{2}. \tag{28b}$$

Noting that the total field E' equals $(E'_x + E'_y)$ we see that the light emergent from the phase plate is either left- or right-circularly polarized light, depending upon whether the phase difference is positive or negative.

If the phase plate is an electro-optic plate subjected to a sinusoidal electric field $V_0 \sin 2\pi ft$, the phase retardation will be equal to $\delta_0 \sin (2\pi ft)$. With V_0 adjusted so that δ_0 equals one-quarter wave retardation, the light emergent from the circular polarizer will oscillate in polarization between lcp and rcp at the frequency f determined by the external field (Velluz et al., 1965). This feature of electro-optic plates, combined with the electronic technique termed phase-sensitive detection, permits sensitive CD measurements, as we will see below.

3. Commercial Circular Dichrometers

(i) Roussel-Jouan

The French firm Roussel-Jouan manufactures an instrument based upon the electro-optic circular polarizer described immediately above and first applied to CD measurements by Grosjean and Legrand (1960). The instrument, whose optical arrangement is given diagramatically in Fig. 9, operates as follows. Light from the deuterium lamp A is rendered

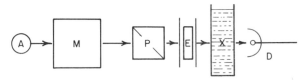

FIG. 9. Optical block diagram of the Roussel-Jouan, Jasco and Cary 60 CD instruments. The symbol A designates the light source; M is the monochromator; P is a linear polarizer; E is an electro-optic plate; X is the sample; and D is the detector.

monochromatic by the single-prism Littrow monochromator M. The monochromatic beam then passes through an electro-optic circular polarizer which consists of a quartz Rochon linear polarizer, P, and an ammonium dihydrogen phosphate EOLM, "E", in series. The EOLM is energized by an alternating voltage V so that the polarization of the beam varies from lcp to rcp at a frequency f. Thus the beam of intensity I at any time t may, to a first approximation, be written as a sum of intensities of lcp and rcp, as follows:

$$I = I_L + I_R$$
$$= I_L^0(1 + \cos 2\pi ft) + I_R^0(1 - \cos 2\pi ft), \qquad (29)$$

where I_L^0 and I_R^0 are equal. The beam then passes through the sample X. If the sample exhibits finite circular dichroism, the absorbances A_L and A_R for lcp and rcp will be different. Now the quantity $(A_L - A_R)$ is generally very small; we designate it ΔA. If the mean of A_L and A_R is designated by A, then, using the series expansion of the exponential function and retaining only the first two terms, the intensity I' of light reaching the photomultiplier D will be

$$I' = I^0\,10^{-A} - 2.303 I^0\,10^{-A}\,\Delta A \cos 2\pi ft. \qquad (30)$$

The first term is large and time-invariant. In the Roussel-Jouan instrument, it is kept constant, regardless of variations in I^0 and in A, by incorporating it as reference in a servo-loop controlling the width of the monochromator slits. The second term, however, varies linearly with ΔA and oscillates in time as $\cos 2\pi ft$. Determination of the magnitude of ΔA requires the separation of this term from the first and also from noise, especially noise originating in the statistical distribution of photoelectrons in the detector. This task is again achieved by phase-sensitive detection of the signal, which in this instance is equivalent to multiplication of the signal by $\cos 2\pi ft$, followed by a time-averaging process. The time average of the term $2.303 I^0\,10^{-A}\,\Delta A \cos^2 2\pi ft$ is simply $2.303 I^0\,10^{-A}\,\Delta A/2$; other terms, as well as noise, average to zero. Thus a sensitive measurement of ΔA is possible. The ratio of the detected and averaged second term to the first is then determined electronically, and the result, which depends only upon ΔA and not upon I_0 or A, is recorded. It should be noted that only a single sample cell and a single light beam are required, in contrast to ordinary absorption measurements. Furthermore, there are no moving parts in the optical train, with the exception of the wavelength and slit control mechanisms.

Improvements to the Grosjean–Legrand design described above have been reported recently (Grosjean and Tari, 1964; Velluz and Legrand, 1965). The single monochromator and polarizer are replaced in the·new design by a synthetic crystal quartz double-monochromator, in which the second monochromator serves also as linear polarizer. This decreases the stray light in the system and extends the lower wavelength limit of 185 mμ.

(ii) *The Japan Spectroscopic Company (Jasco)*

The Jasco ORD-5 instrument is available with CD measuring mode whose simplified optical diagram and electronic technique are identical to those of the Roussel-Jouan instrument depicted in Fig. 9. In this mode, the Jasco instrument uses a fused silica double-prism monochromator, quartz Rochon linear polarizer, and EOLM driven by an alternating field. The Jasco instrument, however, utilizes programmed slits and variable photomultiplier gain to compensate variations in I^0 and in A.

(iii) *Cary Instruments*

Cary Instruments offer a CD attachment, designated Model 1401, for their Model 14 recording spectrophotometers. In addition, an attachment for the Model 60 spectropolarimeter is now available. The 1401 accessory fits into the sample and reference compartments of the spectrophotometer and renders the reference beam right-circularly polarized, the sample beam left-circularly polarized. This is accomplished by ammonium dihydrogen phosphate linear polarizers and Fresnel rhombs, which act as quarter-wave plates. Identical solutions are placed in the sample and reference channels, and the difference in absorbance recorded in the normal way. The device has three advantageous features: it is a simple device; the CD measurement is absolute; and finally, the CD measurement can be calibrated with neutral density filters. The disadvantages of the system are its low precision, poor baseline, and limited wavelength range.

The CD attachment for the Cary 60 spectropolarimeter utilizes the electro-optic phase plate method to generate circularly polarized light. Thus the simplified optical diagram for this attachment and its method of operation are largely identical to that for the Roussel-Jouan instrument above. The precision and wavelength range of this system are much superior to those of the Cary Model 14 attachment.

(iv) *Rehovoth Instrument Company*

The Rehovoth Instrument Company, an Israeli firm, also manufactures a CD measuring attachment for the Cary 14. This system is similar to the 1401 attachment, but uses, instead of the Fresnel rhombs, quartz multi-wave plates with optic axis in the plane of the plate (Holzwarth, 1965). The precision of this system, like that of the 1401 attachment, is at least an order of magnitude poorer than that of the Roussel-Jouan, Jasco and Cary 60 systems.

(v) *Shimadzu*

The Shimadzu Company in Japan markets a CD attachment for their manual single-beam spectrophotometer. The principle is similar to the design of Mitchell (Mitchell, 1957), and is also of much lower precision than the electro-optic methods.

4. METHODS

(a) Calibration of the Polarimeter and Circular Dichrometer

One common means of calibrating the polarimeter is measuring the ORD of a known standard such as a freshly prepared sucrose solution (U.S. National Bureau of Standards grade). Its Standard Sample 17 gives

$$[\alpha]^{20}_{5461\,A} = 78.342 \quad \text{and} \quad [\alpha]^{20}_{5892.5\,A} = 66.529 \tag{31}$$

for a 26% (w/v) sucrose solution.

The $[\alpha]_D$ of sucrose at the sodium D line (589 mμ) is also reported to be

$$[\alpha]^{20}_D = 66.412 + 0.01267w - 0.000376w^2, \tag{32}$$

for $w = 0$–50 g per 100 g solution and

$$\alpha^t = \alpha^{20}[1 - 0.00037(t - 20)], \text{ for } t = 14\text{–}30°C. \tag{33}$$

In practice, such minor variations in rotations with concentration and temperature are usually insignificant. Lowry and Richards in 1924, as quoted by Lowry (1935), found that

$$[\alpha] = 21.648/(\lambda^2 - 0.0213) \tag{34a}$$

for a 26% (w/v) sucrose solution using the ORD data between 671 and 383 mμ, whereas Harris *et al.* (1932) reported that the one-term Drude equation was obeyed as far as 236 mμ with

$$[\alpha] = 21.676/(\lambda^2 - 0.0213). \tag{34b}$$

Another simple means for calibration is to use the quartz control plates, consisting of one left and one right rotating quartz plate engraved with rotations for 589 and 546 mμ with U.S. National Bureau of Standards certificate. The plates can be purchased from O. C. Rudolph & Sons, Inc., Caldwell, N.J., U.S.A.

Unlike the standards available for ORD calibration, no compounds of known CD have been accepted as standards. Velluz *et al.* (1965) have listed the $(\epsilon_L - \epsilon_R)$ of many organic compounds, which can be used for comparison. For example, d-camphor in dioxane has an $(\epsilon_L - \epsilon_R)$ of $+1.6$ at 300 mμ. We do not know how minor variations in concentration and/or temperature of the solution affect CD. As CD measurements become increasingly popular, some standards will eventually be developed and accepted for routine calibrations. (See Note added in proof.)

(b) Difference Spectropolarimetry

Like the difference absorption spectra, difference spectropolarimetry has the advantage of measuring small changes in rotation when the sample is compared with a reference material. In an instrument that has a folded beam, such as the Cary Model 60 (see Section 3 (b), Fig. 5), the differential measurements are made possible by placing a reference cell in the beam after it reflects from the Faraday cell. Thus, the reference cell sees the light beam in a direction opposite to that seen by the sample cell; a right-hand rotator in the reference cell acts like a left-hand one to the instrument and vice versa. If both the sample and reference cells of equal light path are filled with the same optically active solution, the sample rotation is exactly cancelled by the reference rotation, and the baseline is flat. If the sample cell has a rotation slightly different from that of the reference cell, the difference polarimetry can detect such small difference, as the large rotations of the solution are cancelled in this technique. The protein concentration can further be increased by, say, tenfold and any small difference in rotation is ten times what it would otherwise be, provided of course that the total absorbance of each cell does not exceed one. Difference polarimetry is therefore very useful in detecting small conformational changes. Adkins and Yang (1967) found that artifacts occur if the solution used contains absorbing material. For instance, placing two identical solutions of K_2CrO_4 (with absorbance of about 0.5 each) at positions for the sample and reference cells shows false levorotations with a profile similar to the absorption spectrum with a maximum levorotation at the absorption maximum. These artifacts are normally small if the absorbance in the reference cell (with light reflected from the Faraday cell; see Fig. 5) is less than about 0.5, but increase almost exponentially at higher absorbance. (The magnitude of the artifacts, however, may vary with the particular instrument used, perhaps depending on the instrument alignment.) Thus, this technique should present little problem when difference ORD is measured away from any absorption bands. The problem can also be circumvented even in the absorbing region by first measuring the difference rotation of the sample cell against the reference cell and then repeating the measurements through interchange of the sample and reference cells. The artifacts are cancelled by subtracting the second set of readings from the first, and the resultant difference rotation equals twice the actual one. This procedure is permissible if the sample and reference cells have nearly identical absorbance.

Chignell and Gratzer (1966) suggested a simple way to reduce the large rotations of the proteins. They placed the sample cell in series with another cell containing an optically active substance of opposite rotation: for example, a levorotatory protein in series with a dextrorotatory sucrose. Thus most of the large rotation of the protein is cancelled out by the rotation of the sucrose. This method achieves much the same objective as difference spectropolarimetry, but the rotations of the compensating compound must be known beforehand.

(c) Measurements in the Absorbing Region

The subject of precision and accuracy of the measurements has already been discussed in Section 3 (a). Difficulties are commonly encountered for measurements in the absorbing regions. One rule of thumb is to maintain the total absorbance of the solution including the cell walls well below two; that is, at least more than 1% of the intensity of the incident light at the chosen wavelength is transmitted (Urnes and Doty, 1961). The use of a short light path of, say, 0.1 mm or less is often necessary when measurements extend below 200 mμ, where most of the solvents absorb light (the transmittance increases exponentially with decreasing concentration or light path (Eq. (7b)). One simple way to detect an artifact is by varying the concentration of the solution and/or by using cells of different light path. A genuine Cotton effect should remain invariant under these conditions.

Sometimes strongly absorbing materials can be measured in very dilute solutions, which may have rotations of less than $\pm 0.01°$. If so, the baseline of the blank must be determined accurately. With the present recording spectropolarimeters, one sometimes finds that the baseline shifts a few millidegrees during the changes of solution and solvent. This can cause serious errors in the measured rotations of the solution and must be corrected. One such procedure is to compare the specific rotations of the sample near the end of the absorption band with those obtained with a more concentrated solution in a cell of longer path length. The wavelength range of the latter overlaps portions of the absorption region. The baseline of the blank for the Cotton effect region is then slightly adjusted, if necessary, so that the overlapping portions yield identical specific rotations in the two sets of experiments.

5. EXPERIMENTAL OBSERVATIONS

(a) Visible Rotatory Dispersion

Most globular proteins in aqueous solution display levorotation in the visible region; for instance, the specific rotation at sodium D line, $[\alpha]$ is usually between -30 and -70, and decreases to near -100 upon denaturation. This strongly suggests some structural elements are common to the protein molecules. They are optically active and have a positive $[\alpha]_D$ which vanishes with the disruption of such elements. The α-helix of Pauling *et al.* (1951) is one such ordered structure, which can be twisted into two nonsuperposable forms, the right-handed and left-handed helices. (A right-handed helix of an L-polypeptide is the mirror image of a left-handed one of the D-isomer.) If the two kinds of spirals can be formed with equal probability, the rotations from the right-handed and left-handed helices would cancel each other, but the helical backbone would show a rotatory contribution in addition to that due to the asymmetric amino-acid residues, if one-handness prevails. This idea was fully supported by the experimental observations of the ORD of synthetic polypeptides in the middle 1950's. A typical example is poly-γ-benzyl-L-glutamate, which is known to be helical in a poor solvent and randomly coiled in a good solvent (Doty *et al.*, 1954). The helical form always displays an anomalous dispersion; it is dextrorotatory at longer wavelengths, passes through a maximum, decreases sharply and changes the sign of rotation at shorter wavelengths. Once the ordered structure is destroyed, the ORD is levorotatory in the visible region and its magnitude increases monotonically with decreasing wavelength (i.e. a normal dispersion). Figure 10 clearly illustrates the fact that the contributions of the α-helix are in large part responsible for the difference in the ORD profiles of ordered and disordered polypeptides, a finding in full agreement with earlier expectations. Similar results were observed for poly-L-glutamic acid; its conformation in aqueous solution can be controlled simply by adjusting its pH (Doty *et al.*, 1957). Today visible rotatory dispersion remains a most sensitive tool for detecting and estimating the helical contents of proteins in solution, in spite of its many shortcomings (Section 7). [Collagen is a notable exception among the proteins studied; its $[\alpha]_D$ is about -300 in the native state and increases toward -100 upon conversion to gelatin (denaturation). The unusually high proportion of proline and hydroxyproline residues in this protein makes the formation of α-helix unlikely. Collagen is known to consist of three left-handed polypeptide chains twisted together into a right-handed helix, a structure quite different from an α-helix (for a review see Harrington and Von Hippel, 1961).]

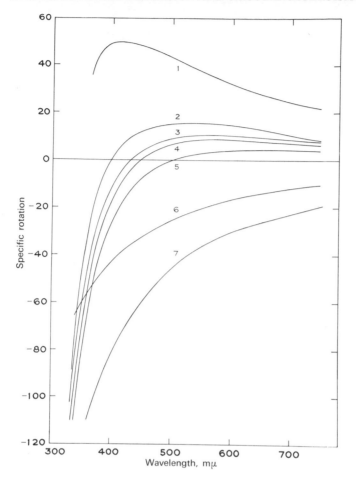

FIG. 10. Optical rotatory dispersion of poly-γ-benzyl-L-glutamate. Helix-promoting solvents: 1, *m*-cresol; 2, chloroform; 3, ethylene dichloride; 4, dioxane; and 5, dimethylformamide. Coil-promoting solvents: 6, dichloroacetic acid; and 7, hydrazine. (Yang and Doty, 1957.)

Many synthetic polypeptides are helical in poor solvents and their ORD substantiates the foregoing conclusions drawn from poly-γ-benzyl-L-glutamate and poly-L-glutamic acid. Among the polymers studied are poly-L-alanine, poly-ε-*N*-carbobenzoxy-L-lysine, poly-L-lysine, poly-L-leucine, poly-L-methionine, poly-γ-methyl-L-glutamate, and various copolymers. These have been summarized in the review by Urnes and Doty (1961).

That the helices of the polypeptides are one-handed can further be demonstrated by experiments on D,L-copolymers. For instance, the random incorporation of D-isomers into an L-polypeptide would proportionally reduce the magnitude of rotations of the asymmetric residues. This is indeed the case when the polypeptide is present in its disordered form (Blout *et al.*, 1957). On the other hand, in a helix-promoting solvent the rotations in the visible region actually rise with the inclusion of a small amount of D-isomer because the helical backbone, which is dextrorotatory at, say, sodium D line, retains its one-handedness, whereas the levorotation of the L-residues is partially cancelled out by the dextrorotation of the D-isomers. If too much D-isomer is introduced, and the ratio $D/(L+D)$ approaches 0.5, the probability of forming helices of opposite screw senses becomes great and eventually meso-helices occur and the rotation becomes zero (see Section 8 (a) on helix-coil transition). Whether the helic formed by L-polypeptides will be right- or left-handed had been controversial. Strong evidence has now come from the X-ray diffraction studies of myoglobin, hemoglobin and lysozyme, the helical segments of which prove to be all right-handed (Kendrew *et al.*, 1961; Perutz *et al.*, 1961; Blake *et al.*, 1965). Save for the remote possibility that the wet crystals of the three proteins unfold and rewind themselves in solution into helices of the opposite sense, we may conclude that the proteins and L-polypeptides favor the formation of right-handed helices. The exclusion diagram of Ramachandran *et al.* (1963) seems also to slightly favor the right-handed α-helix over the left-handed one. (Two adjacent planar amide groups linked through an α-carbon atom form two dihedral angles with the single bonds, $N-C_\alpha$ and $C_\alpha-C'$. Ramachandran and his co-workers calculated the interatomic distances of all important atoms over all range of the dihedral angles which are of interest and selected a set of contact distances for the various kinds of atoms which are fully allowed and another which is considered minimal on the basis of the selected van der Waals' radii for all the atoms concerned. Plotting these pairs of dihedral angles against each other maps out the fully allowed and outer limit regions of the conformational plane, beyond which are the excluded regions.) Exception to this rule, however, is not ruled out, especially since spatial arrangement of the side groups can make it difficult to fold the polypeptide chain into a right-handed helix. The exclusion diagram also suggests that a left-handed α-helix is still within the outer limits of the theoretical calculations. A notable example is poly-β-benzyl-L-aspartate, which forms a left-handed helix, albeit unstable. According to the recent theoretical calculations of Scheraga *et al.*(1967), the dipole interaction of the ester group with the backbone stabilizes the right-handed helix in, say, poly-γ-methyl-L-glutamate, but destabilizes it in poly-β-benzyl-L-aspartate, thus explaining the difference in screw-sense between the two polymers. Poly-L-aspartic acid, however, forms a right-handed helix. The results of ORD studies of synthetic polypeptides were originally thought to agree with a right-handed helix on the basis of the Moffitt theory (1956), but this prediction was discarded after the drastic revision of the earlier theories of helical rotation (Moffitt *et al.*, 1957)(see Section 6 (b)).

Structural elements other than the α-helix are also important, although ORD evidence of their existence is still scarce. Pauling and Corey's parallel and antiparallel pleated sheets or the β-forms (1951) are two examples. A third is the cross-β forms, in which the polypeptide chains fold back on themselves in an antiparallel fashion to form sheetlike structure intra-molecularly (originally, the cross-β form was defined as one in which the chains were oriented perpendicular to the fiber axis, whereas in the parallel and antiparallel β-forms the chains are parallel to the fiber axis). These β-forms are optically active. The first experimental observation came from the ORD results of oligomers of γ-benzyl-L-glutamate; their β-aggregates (based on infrared spectral evidence) display dextrorotatory dispersion in the visible region, which does not resemble that of either the helices or coiled forms. The β-aggregates (Fig. 11) dissociated upon dilution or addition of another polar solvent such as formamide. Thus, their rotations decrease with concentration until

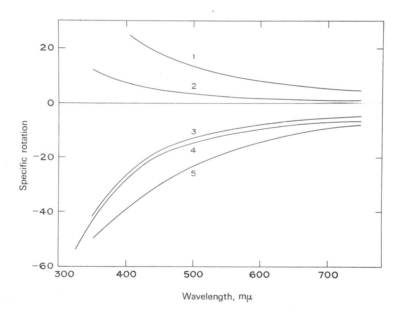

FIG. 11. Optical rotatory dispersion of an oligopeptide of γ-benzyl-L-glutamate at different concentrations. Curves 1 to 3, 6.3, 2.6, and 0.5% in chloroform; 4, 2.6% in chloroform saturated with formamide; and 5, 1% in dichloroacetic acid. (Yang and Doty, 1957.)

finally they become levorotatory, as is the coiled form. Most of the β-aggregates in solution are difficult to reproduce exactly. Furthermore, the conditions that favor the extensive intermolecular β-aggregates tend to cause precipitation. This problem, however, does not arise with intramolecular cross-β forms, which should be concentration-independent. Our knowledge of the properties of β-structures in solutions is still very limited. Among the polypeptides and proteins that are known to produce β-forms as evidenced from their ORD results (in addition to the oligomers of γ-benzyl-L-glutamate mentioned above; see also Wada *et al.*, 1961) are poly-O-acetyl-L-serine (Fasman and Blout, 1960; Imahori and Yahara, 1964), poly-O-benzyl-L-serine (Bradbury *et al.*, 1962), poly-S-carbobenzoxymethyl-L-cysteine (Harrap and Stapleton, 1963; Ikeda *et al.*, 1964; Anufrieva *et al.*, 1965), poly-S-methyl-L-cysteine, poly-S-carbobenzoxy-L-cysteine, poly-S-carbomethyoxyethyl-L-cysteine, and poly-S-carbobenzoxyethyl-L-cysteine (Kamashima, 1966), poly-L-lysine (Sarkar and Doty, 1966; Davidson *et al.*, 1966), denatured bovine serum albumin and ovalbumin (Imahori, 1960), and silk fibroin (Iizuka and Yang, 1966) (see also Section 8 (b) on coil-β and helix-β transitions).

(b) Ultraviolet Rotatory Dispersion

The featureless ORD curves of polypeptides and proteins in the visible region (Figs. 10 and 11) are of course influenced by the optically active absorption bands in the ultraviolet region, since the magnitude of rotations falls off rather slowly on both sides of a Cotton effect (Section 2 (c)). In principle, then a direct observation of the Cotton effects of a helix, β-form, and coil will provide proof of ORD as a powerful technique for studying the conformations of polypeptides and proteins. Simmons and Blout (1960) first detected a trough at 233 mμ for tobacco mosaic virus protein. Subsequently, improved instrumentation extends measurements to about 185 mμ and maps out more details of the Cotton effects. Figure 12 illustrates typical Cotton effects for an α-helix, which displays a 233-mμ trough, a 198-mμ peak, and a shoulder near 215 mμ with a cross-over (zero rotation) near 220 mμ (Yang, 1967b). Most recently, a second cross-over has been found near 186 mμ; another trough is believed to be present near 182–184 mμ (Cary Instruments and Durrum Instrument Corporation, private communications), although the exact position is still uncertain because of instrumental limitations. Other Cotton effects below 180 mμ are still beyond the reach of the present instruments. The broken line in Fig. 12 (and also Fig. 13) is the calculated ORD from CD using the Kronig–Kramers transform (Eq. (16)).

In striking contrast to the α-helix, the coiled form shows a negative Cotton effect with a 205-mμ trough and a 190-mμ peak and another much

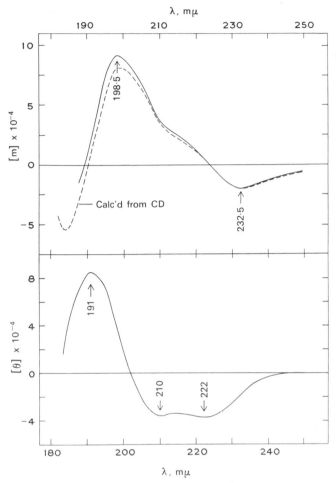

FIG. 12. Mean residue rotation, $[m]$, and mean residue ellipticity, $[\theta]$, of helical poly-L-glutamic acid in water at pH 4.75. Dashed line: calculated $[m]$ from $[\theta]$ using the Kronig–Kramers transform. (Yang, 1967b.)

smaller Cotton effect between 230 and 240 mμ (Fig. 13). Table I summarizes the *known* Cotton effects of the α-helix, coiled form, and the recently observed β-form (Yang, 1967b). The magnitude of the peaks and troughs listed should be regarded as tentative.

The Cotton effects of the β-form are more difficult to measure than are those of the helical and coiled forms, either because most of the β-aggregates are not soluble in any solvent without breaking up the ordered structure or because the mixed organic solvents used absorb strongly in the ultraviolet region. At the time of this writing, however, three

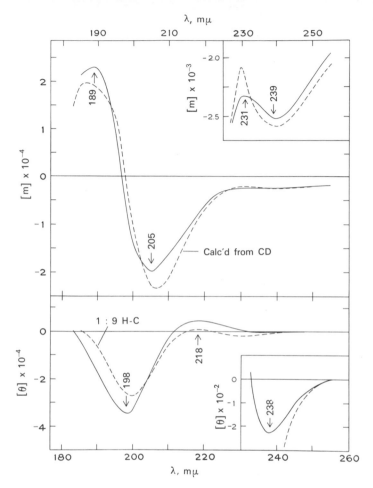

FIG. 13. Mean residue rotation, [m], and mean residue ellipticity, [θ], of coiled poly-L-glutamic acid in water at pH 7.39. Top dashed line: calculated [m] from [θ] using the Kronig–Kramers transform. Bottom dashed line: calculated [θ] of a hypothetical mixture of 1 : 9 helix-coil based on the data in Figs. 12 and 13. (Yang, 1967b.)

publications on the β-form of poly-L-lysine (Sarkar and Doty, 1966; Davidson et al., 1966) and silk fibroin (Iizuka and Yang, 1966) have appeared and seem to give the same results. Poly-L-lysine will be further discussed in Section 8 (a) under α-β transition. Figure 14 illustrates the Cotton effects of silk fibroin in mixed solvents. There is a trough at 229 mμ, a peak at 205 mμ (but with no shoulder near 210 mμ as occurs in the α-helix) and another trough near or below 190 mμ. Table I and Figs. 12 and 14 indicate that the magnitude of the peaks and troughs for the β-

TABLE I. THE OPTICAL ROTATORY DISPERSION AND CIRCULAR DICHROISM
OF VARIOUS CONFORMATIONS*†

$\lambda, m\mu$	α-Helix‡		β-Form§		Coil‖	
ORD	$\lambda, m\mu$	$[m']$	$\lambda, m\mu$	$[m']$	$\lambda, m\mu$	$[m']$
Trough	232.5	−15,000	229–30	(5000)	239	−2,000
Peak	198.5	+68,000	205	(+24,000)	231	−1,800
Trough	182–4	−	~190	(−17,000)	205	−19,000
Peak					189	+16,000
CD	$\lambda, m\mu$	$[\theta]$	$\lambda, m\mu$	$[\theta]$	$\lambda, m\mu$	$[\theta]$
Minimum	222.5	−41,000	218	(−20,000)	238	−300
Minimum	210	−38,000				
Maximum	191.5	+90,000	195–7	(+48,000)	218	+4,900
Minimum					197.5	−38,000

*Taken from Tang (1967b) and revised (see Note added in proof).
†The magnitude of the extrema listed here is still tentative. Dimensions: reduced mean residue rotations, $[m']$, in deg cm²/decimole and mean residue ellipticity, $[\theta]$, in deg cm²/decimole (uncorrected for refractive index).
‡Poly-L-glutamic acid in water at pH 4.75.
§Silk-fibroin in 50% methanol (extrapolated to 100% β-form). The values in the parenthesis are only rough estimates and vary with the solvent composition.
‖Poly-L-glutamic acid in water at pH 7.39.

form is much smaller than that for the α-helix. Thus, the rotations due to the β-form can easily be outnumbered by those of the helices when both conformations coexist in a protein molecule. This is one of the difficulties encountered in the analyses of the ORD data of proteins (see Section 7 (d)).

(c) Circular Dichroism

The Cotton effects, as shown in Figs. 12–14, will of course find the corresponding CD bands over the same wavelength range (Yang, 1967b). Holzwarth and Doty (1965) (see also Holzwarth, 1964) first reported the CD of the helical and coiled conformations. Data on CD of polypeptides and proteins are just beginning to accumulate, now that commercial instruments have become available. The α-helical form has two negative dichroic bands with a double minimum at 222 and 210 mμ and a large positive band near 191 mμ. The coiled form has a large negative band at 198 mμ, a positive one at 218 mμ, and another very small negative band at 238 mμ (the last one easily escapes detection unless the concentration of the polymer is high enough to increase the intensity of the dichroism). Two CD bands are observed for the β-form, probably of the antiparallel type, one negative at 218 mμ and the other positive at 197 mμ (Sarkar and Doty, 1966; Townend et al., 1966; Iizuka and Yang, 1966). Table I also includes the CD of the three conformations. Here again, the magnitude of extrema should be considered tentative.

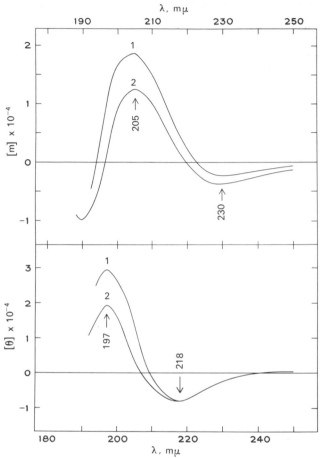

FIG. 14. Mean residue rotation, $[m]$, and mean residue ellipticity, $[\theta]$, of the β-form of silk fibroin in methanol–water. Methanol (v/v): 1, 93%; and 2, 50%. (Iizuka and Yang, to be published.)

On the basis of exciton theory, Moffitt (1956) first predicted that the amide π–π^* transition in an α-helix would be split into two absorption bands of opposite polarization and that strong rotatory bands of equal intensity but opposite sign would be associated with the parallelly and perpendicularly polarized bands. Subsequent refinement of this treatment by Tinoco and his co-workers (Tinoco, 1961; Tinoco et al., 1963; Bradley et al., 1963; Woody, 1962; Tinoco, 1964) indicated four rotatory bands rather than two for the 190-mμ amide π–π^* transition. These, according to Woody (1962), should occur at 185, 189, 193 and 195 mμ. In addition, Schellman and Oriel (1962) and Woody (1962) predicted an n–π^* transition of the amide group located near 225 mμ, which should contribute significantly to the optical activity of the α-helix. Thus the helix should exhibit as many as five circular dichroic bands between 185 and 230 mμ, a region which is currently attainable with recording circular dichro-meters and spectropolarimeters. These bands of helical polypeptides to date are, however, too close to permit experimentally either an un-ambiguous resolution of the relative roles of the n–π^* and π–π^* transi-tions or a verification of the four separate π–π^* bands as predicted. But polarized absorption spectra of oriented film of helical polypeptide show the presence of two strong bands, one perpendicular at 191 mμ and other parallel at 206 mμ, which are assigned to two of the π–π^* exciton bands predicted by Moffitt (Gratzer *et al.*, 1961; Holzwarth and Doty, 1965). In addition, the polarized spectra give evidence for a weak, perpen-dicularly polarized band at about 222 mμ, which is most probably the amide n–π^* transition. These correlate well with the observed CD bands in Fig. 12, which displays three extrema at 191, 210 and 222 mμ. Holz-warth and Doty (1965) also reported that the observed rotational strength of the above-mentioned three CD bands (assuming they are approximately Gaussian) agreed in sign and order of magnitude with the theoretical predictions of Moffitt, Tinoco, Woody, Schellman and Oriel. These findings provide strong evidence for the involvement of both the n–π^* and π–π^* transitions in the optical activity of helical polypeptides. Holzwarth and Doty (1965) further showed that the rotatory contribu-tions of the helical backbone in the visible region and the corresponding Moffitt parameters (Section 6 (b)) were largely a result of the peptide transitions between 185 and 230 mμ (for more details see the review by Holzwarth, 1964; see also Note added in proof).

The detailed origins of the optical activity of the disordered polymer are still not known, but the strong negative CD band near 198 mμ and the weak positive band near 218 mμ (Fig. 13) probably arise from the amide π–π^* and n–π^* transitions (Holzwarth and Doty, 1965). Neither is it clear whether the newly found, exceedingly weak minimum near 240 mμ is a result of overlapping CD bands of opposite sign or a new band arising from unsuspected electronic transitions. We must wait for further experimental and theoretical investigations on the disordered polymers.

The theoretical calculations of optical activity of the β-form are just beginning (Pysh, 1966; Woody, private communication). Pysh con-sidered the contributions of exciton, coupled oscillation, and one-electron interactions to the rotational strengths of the β-pleated sheets and con-cluded that in contrast to α-helix, the exciton contribution in these struc-tures is not large. Evaluation of the complete expression for rotational strengths of the anti-parallel β-form leads to a positive CD band at 195–198 mμ for the π–π^* transition and a small negative band near

218 mμ for the n–π^* transition. Qualitatively, these findings agree well with experimental observations of the β-form of silk fibroin (Fig. 14) and poly-L-lysine (Sarkar and Doty, 1966; Davidson *et al.*, 1966). Pysh also predicted a large π–π^* Cotton effect at 181 mμ for the parallel β-form, which is sufficiently different from that in the antiparallel β-form to allow differentiation between the two forms. The parallel β-form is also expected to have a net small, negative Cotton effect near 216 mμ. At present experimental confirmation of the parallel β-form in solution is lacking.

(d) Cotton Effects of Nonpeptide Chromophores

The Cotton effects and CD of proteins are of course not confined to peptide linkages. Any compound that absorbs may become optically active when it is attached to the proteins or polypeptides. The group includes the aromatic groups in the side chains, a prosthetic group such as heme, or any compound having a chromophore bound to the protein molecule. For instance, the heme proteins, myoglobin, hemoglobin, catalase and cytochrome C all show distinctive Cotton effects in the Soret band (Beychok and Blout, 1961; Yang and Samejima, 1963; Samejima and Yang, 1964; Breslow *et al.*, 1965; Urnes, 1965; Urry and Doty, 1965). Furthermore, they also depend on the ligands attached to the heme group, although no significant difference in the secondary structure of the protein moiety can be detected for these complexes (ferri-, ferro-, carbonyl-, oxy-, and cyan-myoglobin of sperm whale) (Samejima and Yang, 1964). Recently, T. Samejima (private communication) reinvestigated the ORD and CD of ferri-, ferro- and cyan-myoglobin of horse heart and reached essentially the same conclusions (Figs. 15 and 16). Beychok (1967) also reported that poly-L-histidine can produce CD bands when the synthetic polymer is complexed with hemin. Removal of the heme group from hemoglobin or myoglobin immediately destroys the Cotton effects in the Soret bands (Harrison and Blout, 1965; Breslow *et al.*, 1965).

Synthetic polypeptides having chromophores in the side groups, such as tyrosine, tryptophan and phenylalanine, produce Cotton effects in their corresponding absorption bands which overlap those of the peptide chromophores (Fasman *et al.*, 1964; Beychok and Fasman, 1964; Fasman *et al.*, 1965; Sage and Fasman, 1965). Until very recently, however, few Cotton effects associated with these aromatic groups in proteins have been observed. A striking example is the ORD of human carbonic anhydrase which displays several small but distinctive Cotton effects in the region between 260 and 300 mμ (Fig. 17). Among other proteins whose nonpeptide Cotton effects and/or CD have been observed are chymotrypsinogen, chymotrypsin, trypsinogen, trypsin, pepsinogen,

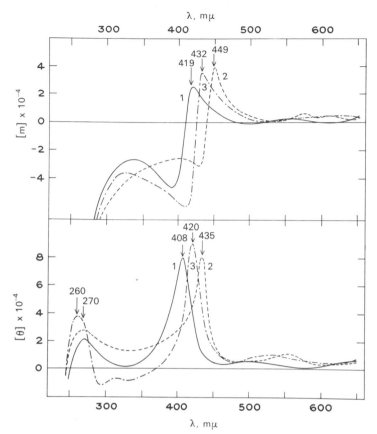

FIG. 15. Mean residue rotation, $[m]$, and mean residue ellipticity, $[\theta]$, of horse heart myoglobin in the Soret band. Curves 1 to 3, ferri-, ferro-, and cyan-myoglobin. Solvent: 0.1 M phosphate buffer (pH 7.0). (Courtesy of T. Samejima, unpublished work.)

pepsin, avidin, strepavidin, bovine serum albumin, β-lactoglobulin (see a brief review on CD of these proteins by Beychok, 1966), and of conalbumin (Tomimatsu and Gaffield, 1965). In most cases these CD's are usually very small and the Cotton effects show only small deviations from the smooth ORD curves, unlike the fine structure of human carbonic anhydrase in Fig. 17. Aside from the aromatic groups, the disulfide linkages of the cystine residues can also be optically active near and below 250 mμ (in a simple aliphatic disulfide the dihedral angle between the planes defined by the two halves of the molecule is close to 90° and the disulfide can exist in one of two rotamer forms, either a right-handed or left-handed screw). The assignment of these nonpeptide chromophores is very much uncertain (see Beychok, 1967).

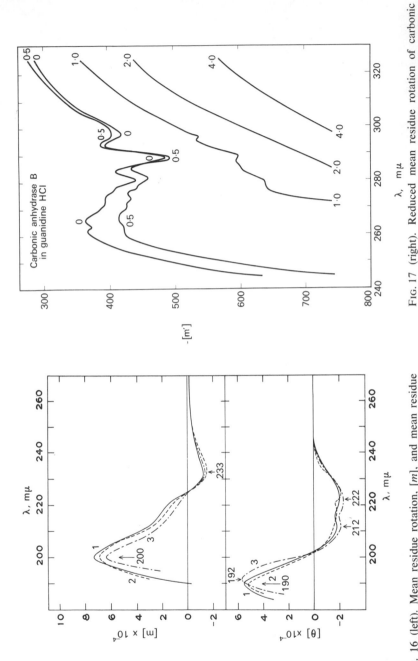

FIG. 17 (right). Reduced mean residue rotation of carbonic anhydrase B in 0.1 M phosphate buffer (pH 7.0). The numbers adjoining the curves indicate the molar concentrations of guanidine hydrochloride. (Myers and Edsall, 1965.)

FIG. 16 (left). Mean residue rotation, [m], and mean residue ellipticity, [θ], of horse heart myoglobin in the ultraviolet region. Symbols, same as in Fig. 15. Solvent: 0.1 M phosphate buffer (pH 7.0). (Courtesy of T. Samejima, unpublished work.)

Small changes in the aromatic Cotton effect of lysozyme have recently been observed by means of difference spectropolarimetry when this protein is complexed with an inhibitor (Adkins and Yang, 1967). Similar changes in the CD for this protein were also reported by Glazer and Simmons (1966). The Cotton effects of nonpeptide chromophores are just beginning to accumulate, and with recording spectropolarimeters and circular dichrometers, many proteins may exhibit such nonpeptide optical activity heretofore undetected. At this writing it is also too early to describe how they will modify the ORD in the visible region.

6. PHENOMENOLOGICAL EQUATIONS

Of all the equations developed in the field of ORD, perhaps the Drude equation is most often mentioned. It can be applied to all optically active compounds, including proteins and polypeptides. Another widely used equation is that proposed by the late Wm. Moffitt (1956) for the ORD of an α-helix. The Moffitt equation is used in estimating the helicity in the protein molecules. In this section we will only show the relationship between the two equations from a phenomenological viewpoint.

(a) The Simple Drude Equation

In 1896 Drude first attempted a theoretical treatment of ORD and deduced a general expression for rotations distant from any of the optically active absorption bands:

$$[\alpha] = \Sigma k_i/(\lambda^2 - \lambda_i^2), \tag{35}$$

where λ_i is the wave length of the ith optically active electronic transition, and k_i a constant. Ironically, the physical model chosen by Drude was later shown to be optically inactive and, further, his mathematical treatment actually did not correspond to his model. Equation (35), however, was later shown to be correct from quantum mechanical considerations. The expression of Rosenfeld (1928), for example, can be written as (cf. Eq. (23a)):

$$[m'] \equiv [m][3/(n^2+2)] = (96\pi N/hc)\Sigma R_i\lambda_i^2/(\lambda^2 - \lambda_i^2) \tag{36a}$$

or

$$[m'] = \Sigma a_i\lambda_i^2/(\lambda^2 - \lambda_i^2), \tag{36b}$$

where N is the Avogadro's number, h the Planck's constant, c the speed of light, R_i the rotational strength of the ith electronic transition and $a_i = 0.915 \times 10^{42}R_i$. Note that Rosenfeld introduced the Lorentz correction into his derivation (Eq. (36a)); thus, for the ORD of each CD band Eq. (36b) differs from Eq. (23b) by a factor of $(n^2+2)/3$.

By expanding Eq. (35) into a Taylor's series in terms of $(\lambda^2-\lambda_c^2)^{-1}$ (λ_c is a parameter to be determined), we have

$$[\alpha] = \Sigma k_i/(\lambda^2-\lambda_c^2) + \Sigma k_i(\lambda_i^2-\lambda_c^2)/(\lambda^2-\lambda_c^2)^2 + O[(\lambda^2-\lambda_c^2)^{-3}]. \quad (37)$$

If $|\lambda_i^2-\lambda_c^2| \ll |\lambda^2-\lambda_i^2|$, we can neglect all except the first term on the right side of Eq. (37). Thus, Eq. (37) reduces to a one-term Drude equation:

$$[\alpha] = k/(\lambda^2-\lambda_c^2) \qquad (38)$$

with $k = \Sigma k_i$. The λ_c so determined has no physical meaning; it does not correspond to any single CD band. The ORD of almost all proteins obeys Eq. (38) extremely well. Furthermore, in most cases λ_c decreases upon denaturation of the proteins (see Section 7 (a)).

We have already shown in Section 2 (c) that each Cotton effect can be approximated into a Drude term in regions distant from the (Gaussian) CD band, but Eq. (38) is only the consequence of a mathematical approximation. Indeed, the sum rule derivable from the theory of optical rotation leads to (cf. Eq. (36 a, b)):

$$\Sigma R_i = 0 \quad \text{or} \quad \Sigma a_i = 0. \qquad (39)$$

Any one-term Drude equation obviously violates this rule. Even when the ORD of an optically active compound (away from the CD bands) cannot fit with one Drude term but with two terms:

$$[\alpha] = k_1/(\lambda^2-\lambda_1^2) + k_2/(\lambda^2-\lambda_2^2) \qquad (40a)$$
or
$$[m'] = a_1\lambda_1^2/(\lambda^2-\lambda_1^2) + a_2\lambda_2^2/(\lambda^2-\lambda_2^2), \qquad (40b)$$

the λ_1 and λ_2 do not necessarily represent two CD bands unless *a priori* $(a_1+a_2) = 0$. Only in special cases where one strong CD band has rotations that overshadow those due to other CD bands, the λ_c in Eq. (38) thus obtained may be taken to be roughly equivalent to the CD extremum. The same would be true for two strong CD bands in Eq. (40b), but these are rare exceptions rather than a general rule. Failure to recognize the sum rule has sometimes led to erroneous identification of the CD band and the interpretation of the ORD data.

The graphical solution of the one-term Drude equation is now commonly done by rearranging Eq. (38) into

$$[\alpha]\lambda^2 = \lambda_c^2[\alpha] + k. \qquad (41)$$

Thus, a plot of $[\alpha]\lambda^2$ versus $[\alpha]$ yields a straight line with k and λ_c^2 as the intercept on the ordinate and the slope (Yang and Doty, 1957). The empirical nature of Eq. (38) limits its applicability to a certain range of wavelengths, and it is advisable to report such range in the analysis of the ORD data. The two-term Drude equation with four parameters is more laborious and less unambiguous to solve, although Lowry (1935) has described an algebraic solution of four simultaneous equations with data at four chosen wavelengths. Alternatively, we can rearrange Eq. (40b) into

$$[m](\lambda^2/\lambda_1^2 - 1) = a_1 + a_2(\lambda^2/\lambda_1^2 - 1)/(\lambda^2/\lambda_2^2 - 1). \qquad (42)$$

A value is postulated for λ_1, and a series of curves of $[m](\lambda^2/\lambda_1^2 - 1)$ versus $(\lambda^2/\lambda_1^2 - 1)/(\lambda^2/\lambda_2^2 - 1)$ is plotted with trial λ_2 values until a straight line is obtained or until the chosen λ_1 is discarded because it will not yield a straight line. Next, the λ_1 value is varied and the trial-and-error procedure repeated until the best pair of λ_1 and λ_2 for the data under consideration is found. Such calculations can of course be simplified by using a computer, but the solution is often not unique.

(b) The Moffitt Equation

The ORD of helical polypeptides such as that illustrated in Fig. 10 cannot be fitted with a one-term Drude equation, but can be fitted with two terms, e.g. Eq. (40b). The use of Eq. (40b) was soon discarded, partly because of the uncertainty in solving the four parameters in a two-term Drude equation and, more importantly, because of the introduction of theoretical equation by Moffit (1956). Moffitt's equation for the ORD of an α-helix (Moffitt and Yang, 1956):

$$[m'] = a_0\lambda_0^2/(\lambda^2 - \lambda_0^2) + b_0\lambda_0^4/(\lambda^2 - \lambda_0^2)^2 \qquad (43)$$

is now regarded as empirical because his theoretical treatment neglected several important terms (Moffitt et al., 1957; see also Section 5 (c)). Nevertheless, this equation has gained wide acceptance and, in spite of its imperfection, provides very useful information about the amount of helicity in the protein molecules (see Section 7 (a)). The first term on the right side of Eq. (43) is simply a Drude term. We have already shown the Taylor's series expansion of the general Drude equation (Eq. (37)) and the Moffitt equation differs from the one-term Drude equation (Eq. (38)) only in that the second term of the Taylor's series is non-vanishing for the α-helix (neglecting the higher terms, which are expected to converge quite rapidly if $\lambda \gg \lambda_0$).

Moffitt proposed that the amide absorption bands responsible for the helical rotations are split into perpendicular and parallel (to the helical axis) components. Thus, the Cotton effects of these bands reduce to many Drude terms in the visible region. The partial effective monomer rotation due to such a band pair may be written as

$$[m_i'] = a_{\parallel}\lambda_{\parallel i}^2/(\lambda^2 - \lambda_{\parallel i}^2) + a_{\perp}\lambda_{\perp i}^2/(\lambda^2 - \lambda_{\perp i}^2)$$

$$\cong a_i\lambda_i^2/(\lambda^2 - \lambda_i^2) + b_i\lambda_i^4/(\lambda^2 - \lambda_i^2)^2. \tag{44}$$

Moffitt further indicated that the rotational strength, which is proportional to a_i, of each parallel component may be exceedingly large but almost exactly compensated by the opposing rotational strength of its perpendicular partner, i.e. $a_{\parallel i} \cong -a_{\perp i}$. It can then easily be shown (Iizuka and Yang, 1965) that λ_i is in between $\lambda_{\parallel i}$ and $\lambda_{\perp i}$ and that b_i in Eq. (44) will not vanish. Formally, we have

$$[m'] = \Sigma a_i\lambda_i^2/(\lambda^2 - \lambda_i^2) + \Sigma b_i\lambda_i^4/(\lambda^2 - \lambda_i^2)^2 \tag{45}$$

which is identical with Eq. (42), provided that

$$a_0\lambda_0^2 = \Sigma a_i\lambda_i^2, \quad b_0\lambda_0^4 = \Sigma b_i\lambda_i^4$$

and

$$b_0\lambda_0^6 = \Sigma b_i\lambda_i^6 \tag{46}$$

(assuming $|\lambda_i^2 - \lambda_0^2| < \lambda_i^2$) (Moffitt and Yang, 1956).

The parameters a_0, b_0 and λ_0 can be determined graphically by rearranging Eq. (43) into

$$[m'](\lambda^2 - \lambda_0^2) = a_0\lambda_0^2 + b_0\lambda_0^4/(\lambda^2 - \lambda_0^2) \tag{47a}$$

or

$$[m'](\lambda^2/\lambda_0^2 - 1) = a_0 + b_0/(\lambda^2/\lambda_0^2 - 1) \tag{47b}$$

and by plotting the left side against the reciprocal of the denominator of the second term on the right side. A straight line will result when λ_0 is correct (Fig. 18). At present, $\lambda_0 = 212 \text{ m}\mu$ is accepted by almost all workers; its uncertainty was originally set at $\pm 5 \text{ m}\mu$, but more recently it was reduced to $\pm 2 \text{ m}\mu$ by a computer program first proposed by Sogami et al. (1963). The numerical values of $(\lambda^2/\lambda_0^2 - 1)$ and $1/(\lambda^2/\lambda_0^2 - 1)$ at various wavelengths are listed in the Appendix.

The early experimental data on poly-γ-benzyl-L-glutamate (Fig. 10) and poly-L-glutamic acid in several helix-promoting solvents can be fitted with the Moffitt equation (between 300 and 700 mμ). The average b_0 is -630 but the a_0 value varies with the solvent used. This conclusion was later confirmed with other polypeptides having different amino acid side

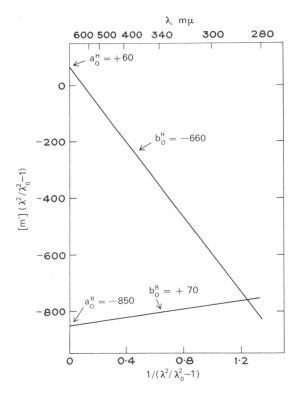

FIG. 18. Graphical representation of the Moffitt equation for the dispersions of helical (H) and coiled (R) poly-L-glutamic acid.

groups. Exceptions do occur but can be explained by various factors such as the opposite handedness of the helices and a positive nonhelical b_0 contribution arising from the nonpeptide chromophores (Yang, 1967a). They actually support the contention that a large negative b_0 indicates the presence of a right-handed helix (Section 7 (c)). A new complication comes from the recent work of Cassim and Taylor (1965), who found that b_0 is dependent on the refractive index of the solvent used even though the polypeptide is completely helical (see Section 7 (c)).

(c) The Two-term Drude Equation

The observation of apparently two Cotton effects (between 190 and 240 mμ) for the helical form (Fig. 12) has renewed interest in the two-term Drude equation for visible rotatory dispersion of polypeptides and proteins (cf. Eq. (40b)). Imahori (1963) first reported that the ORD of helical poly-γ-benzyl-L-glutamate fit well with Eq. (40b) using $\lambda_1 = 190$

$m\mu$ and $\lambda_2 = 220$ mμ. Later, Yamaoka (1964) found that 193 and 226 mμ for λ_1 and λ_2 would yield a straight line for Eq. (40b). Shechter and Blout (1964a) in turn suggested the use of 193 and 225 mμ. Unlike the Moffitt equation which has a single λ_0 value over a chosen range of wavelength, numerous pairs of λ_1 and λ_2 for Eq. (40b) have been shown to fit the experimental data; for example, $\lambda_1 = 10$ mμ and $\lambda_2 = 250$ mμ or $\lambda_1 = 211$ mμ and $\lambda_2 = 212$ mμ for poly-γ-benzyl-L-glutamate in dimethyl-formamide for wavelengths between 300 and 600 mμ (Yang, 1967a).

In an attempt to reproduce the Cotton effects (Fig. 12), Yamaoka (1964) has further modified the two-term Drude equation by including the damping factors, Γ's:

$$[m'] = \Sigma a_i \lambda_i^2 / [(\lambda^2 - \lambda_i^2) + \Gamma_i^2 \lambda^2 / (\lambda^2 - \lambda_i^2)] \qquad (48)$$

With $i = 193$ and 225 mμ several trial values for Γ_i were used, but none gave satisfactory agreement between Eq. (48) and the experimental curve. Yamaoka proposed that a third term be added to Eq. (48).

(d) Calculation of the Rotational Strength

The rotational strength of a CD band can be calculated from Eq. (20) either graphically or with the aid of a computer program. In the absence of CD measurements we can still estimate the rotational strength by assuming a Gaussian form for the CD band and utilizing the Kronig–Kramers transform. According to Eq. (19), the integral times the exponential in the bracket reaches its extrema at $c = \pm 0.93$ (see Table AIII in the Appendix). Thus, the band width becomes

$$\Delta = |\lambda_{peak} - \lambda_{trough}|/1.86. \qquad (49)$$

Likewise, the CD extremum, λ_i^0, can be taken as the average of the wavelengths at the peak and trough, i.e. $\lambda_i^0 = (\lambda_{peak} + \lambda_{trough})/2$. With λ_i^0 and Δ_i^0 known and using the absolute molar rotation at the peak or trough (or $|[m]_{peak} - [m]_{trough}|/2$), the rotational strength, R_i, can be estimated from Eq. (22), neglecting the term $\Delta_i^0/2(\lambda + \lambda_i^0)$ which is usually very small. The sign of R_i is the same as that of the Cotton effect curve.

7. ESTIMATION OF THE SECONDARY STRUCTURES IN PROTEINS

Several ORD methods have been proposed for estimating the helical contents of proteins; all are based on the reference values obtained from synthetic polypeptides. The justification for such assimilation derives from the similarities between proteins and synthetic polypeptides. Both are made of L-amino acid residues and have the same α-amide linkages. Thus, their ORD is very similar; for instance, the visible rotatory dispersion of native proteins becomes more negative upon denaturation in the same direction as the conversion of dextrorotations of the helices to levorotations of the random coils. Furthermore, the rotations of the disordered form are very close for both proteins and polypeptides after taking the nature of the side groups into consideration. Unlike the helical polypeptides as shown in Fig. 10, however, most proteins in their native state display only simple, not anomalous, dispersion. A working hypothesis (Yang and Doty, 1957) was thus proposed to explain this difference between proteins and synthetic polypeptides: the protein molecule contains only partial helices and the rest of the molecule is irregular in the sense that they may be effectively regarded as the disordered form similar to that found in synthetic polypeptide. This irregular portion may be just as rigid as that in the helical region, but its pattern is nonperiodic. This difference in randomness between proteins and synthetic polypeptides, however, will be manifest in the quantitative analyses of ORD (Section 7 (c)).

Assuming that the rotatory contributions are additive, any rotatory parameter, P, such as b_0 in the Moffitt equation, can be expressed as

$$P = f_\alpha P^\alpha + f_\beta P^\beta + f_c P^c + \ldots \tag{50}$$

where the P's with superscripts represent the reference values for α-helix, β-form, coil, etc., and the f's are the corresponding fractions in the molecule. The current practice is to assume only the helical and coiled forms are present in the protein molecules. This simplifies Eq. (50) into

$$P = f_\alpha P^\alpha + (1 - f_\alpha) P^c \tag{51a}$$

or

$$f_\alpha = (P - P^\alpha)/(P^\alpha - P^c). \tag{51b}$$

Obviously, this approximation is uncertain if secondary structures other than the α-helix coexist in the molecule.

(a) The b_0 Method

The Moffitt equation has since its introduction been widely used in estimating the helical contents in proteins. (This equation cannot be directly applied to certain conjugated proteins, e.g. nucleo- and mucoproteins, the nonprotein components of which are optically active.) The average b_0 value in Eq. (43) was found to be about -630 for a helix and close to zero for the disordered form. According to Eq. (51b) we then have

$$\text{Fraction of helix, } f_\alpha = -b_0/630. \tag{52}$$

Here $b_0^\alpha = -630$ is assumed to represent the rotatory contribution of the helical backbone only. In the middle 1950's evidence supporting the use of Eq. (52) was only suggestive, as the three-dimensional structures of myoglobin and hemoglobin were still not known then. As Crick and Kendrew (1957) pointed out: "It leads to a strong presumption that some sort of helical configuration is present, and of a single hand, but as far as we know does not discriminate between the α-helix and other helical configurations of a similar kind which have from time to time been proposed." Nevertheless, Crick and Kendrew also remarked that "there is an encouraging parallelism between the X-ray and optical results." For example, the X-ray results suggest a rather high percentage of α-helix for tropomyosin and an unusually small amount of α-helix or other folded structure for ribonuclease, and similarly the b_0 values put tropomyosin (-600) at the top and ribonuclease (-100) well down the list of helical contents. It is against such background that the b_0 method becomes a powerful tool for characterizing the conformations of proteins in solution.

The justification for introducing pseudocomplexity of the ORD data with the Moffitt equation even when the proteins obey the simple one-term Drude equation is best illustrated in Fig. 19, where the ORD of a series of hypothetical mixtures of helices and coils is plotted according to the one-term Drude equation (Section 6 (a)). The data for helical content up to about 40% will within experimental errors satisfy both the Moffitt and Drude equation. This criterion strongly suggests that the mojority of proteins which obey the simple Drude equation would have less than 50% helical contents. Downie, as quoted by Todd (1960) in volume 2 of this series, found that

$$\lambda_c^2 = \lambda_0^2[(\lambda^2 - \lambda_0^2) + (b_0/a_0)(\lambda^2 + \lambda_0^2)]/[(\lambda^2 - \lambda_0^2) + 2(b_0/a_0)\lambda_0^2] \tag{53a}$$

which reduces to

$$\lambda_c \cong \lambda_0(1 + b_0/a_0)^{1/2} \tag{53b}$$

when $\lambda \gg \lambda_0$. The $[\alpha]\lambda^2$ versus $[\alpha]$ plot is essentially linear, i.e. λ_c is constant, over the wavelength range of 300–600 mμ for b_0/a_0 ratios as high as $+0.6$ and as low as -0.2 with λ_0 preset at 212 mμ (Urnes and Doty, 1961).

We have already mentioned that the denatured proteins are more levorotatory (in the visible region) than are the native ones (except collagen). Linderstrøm-Lang and Schellman (1954) further observed that

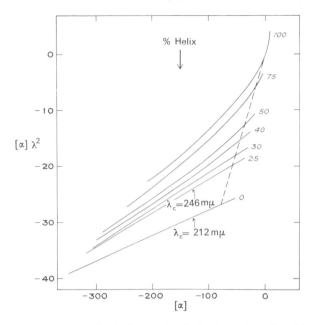

FIG. 19. Dispersion plots for helical and coiled poly-L-glutamic acid together with their hypothetical mixtures according to the simple Drude equation. The dashed line indicates values at 589 mμ. (Yang and Doty, 1957.)

the λ_c of the Drude equation (Eq. 38) for proteins decreased if the protein was denatured. The explanation for these findings can be seen in Fig. 19, where both $[\alpha]_D$ and λ_c increase with increasing helical content. Thus, these two experimental quantities can be used as an indicator of the presence of helicity, but they are only of historical interest and now rarely used, as the b_0 method is far superior to either of them. With spectropolarimeters it is not necessary to limit measurements to a single wavelength. Exceptions to the observations of Linderstrøm-Lang and Schellman have also been found in certain proteins, the λ_c of which either does not drop or increases slightly upon denaturation, but all denatured proteins do reach a common range of between 200 and 220 mμ (Jirgensons, 1961). Equation (53b) indicates that λ_c is usually larger than 212 mμ, as most proteins have negative b_0 and a_0. Furthermore, the higher the helical content, the more negative the b_0 but the less negative the a_0, and therefore the larger the λ_c. If, in rare cases, a_0 and b_0 are of opposite sign, λ_c can then be smaller than 212 mμ. Unlike the b_0 method, both $[\alpha]_D$ and λ_c are dependent on not only b_0 but a_0 as well, the latter being very sensitive to such things as the solvent effect. In view of these uncertainties, the quantitative interpretations of the $[\alpha]_D$ and λ_c methods are very limited.

Instead of using the Moffitt equation, Shechter and Blout (1964 a, b) and Shechter *et al.* (1964) proposed a four-term Drude equation for estimating the helical content of proteins, two terms for the helices and two for the disordered form. What they actually applied was Eq. (40b) using $\lambda_1 = 193$ mμ and $\lambda_2 = 225$ mμ, which is designated as MTTDE (Modified two-term Drude equation). These authors suggested that a_{193} and a_{225} provided two estimates of the helical content. Using the data of poly-L-glutamic acid at pH 4 and 7 as reference states for the helical and disordered forms, they proposed (cf. Eq. (51b)):

$$\text{Fraction of helix} = (a_{193} + 750)/3650 \qquad (54a)$$

and

$$\text{Fraction of helix} = -(a_{225} + 60)/1990 \qquad (54b)$$

for proteins and polypeptides in aqueous solutions. They further found that

$$a_{225} = -0.55a_{193} - 430. \qquad (55)$$

Another set of equations similar to Eqs. (54 a, b) and (55), but with different coefficients, must be used for nonaqueous solutions. To circumvent the solvent dependence of the a-parameters, these authors suggested another equation by taking the average of the two sets of equations (for aqueous and nonaqueous solutions):

$$\text{Fraction of helix} = (a_{193} - a_{225} + 650)/5580. \qquad (56)$$

The many problems associated with the use of MTTDE have been discussed by Yang (1965, 1967a)(see also the review by Harrington *et al.*, 1966). Carver *et al.* (1966b) have modified some of the early conclusions concerning MTTDE and concluded that b_0 as well as $(a_{193} - a_{225})$ best meets the basic assumptions for helical content estimation.

(b) The Trough Method

The direct observation of a 233-mμ trough and a 198-mμ peak is a good indication of the presence of the α-helix in a protein molecule and initially appeared to be superior to the b_0 method for estimating the helical contents. The requirement of minute quantities of proteins for measuring the Cotton effects is also a distinct advantage, especially when the samples are scarce. However, the quantitative aspects of the trough method still leave much to be desired, mainly because the reference values for helices and coils are so uncertain. Not only does the magnitude of the trough vary among the helical polypeptides having different side groups, but different workers found different values for the same polypeptide, such as poly-L-glutamic acid (Yang and McCabe, 1965). The latest $[m']_{233}$ value for helical poly-L-glutamic acid at pH 4.75 is about $-15,000$. Below pH 4.5, aggregation (and eventually precipitation) occurs and the magnitude of the trough increases by more than 10% (Tomimatsu et al., 1966) together with a small red shift of about 1 mμ (Cassim and Yang, 1966). The corresponding coiled form for poly-L-glutamic acid at neutral pH has an $[m']_{233}$ of about -2000. Tentatively, we may write (cf. Eq. (51b)):

$$\text{Fraction of helix} = -([m']_{233} + 2000)/13,000. \qquad (57)$$

A form similar to Eq. (54) can be written for the 198-mμ peak. This method is not widely used, even though the peak magnitude for the helices is several times larger than the trough magnitude, because measurements below 200 mμ are usually very noisy and demand extreme caution. As in the trough method, the reference values at the peak are not thoroughly investigated.

Another problem in using the trough or peak method is the overlapping of nonpeptide Cotton effects. The aromatic groups and disulfide bonds in a protein molecule can produce Cotton effects in their absorption bands and modify both the magnitude and position of the trough and peak (Section 5 (d)). Recently, cyclic amides were reported to have Cotton effects similar to those for the helices (Litman and Schellman, 1965; Balasubramanian and Wetlaufer, 1967). More striking is the ORD of the cyclic decapetides, tyrocidine and gramicidine S–A, which shows a large negative trough near 230 mμ. Gramicidine S–A has an $[\alpha]_{233}$ of about $-16,000$, which is nearly identical to that for helical poly-L-glutamic acid (Ruttenberg et al., 1965). Thus, the trough method, for instance, cannot be regarded as a unique feature of the helices. These cyclic peptides, if present extensively in a protein molecule, could complicate the analyses of the helical content by the trough method.

Aside from the estimation of helical content, the 233-mμ trough can be used to great advantage in the kinetic studies of protein denaturation. Any drastic decrease in helical content would be reflected in the reduction of the trough magnitude. Very few experiments, however, have used this approach.

What we have discussed about the Cotton effects can equally well be applied to the CD bands (Fig. 12). The extrema of these bands or the CD profile as a whole can provide additional measures of helical content. The technique is just beginning to receive attention, and presently very few such calculations have been reported.

(c) Some Problems in Phenomenological Treatments

The very fact that the Moffitt equation is now empirical cautions us against too literal interpretation of its phenomenological treatments. Moffitt considered only the rotations of the helical backbone, not the side groups, in his treatment. He did, however, suggest that a_0, not b_0, in Eq. (43) is sensitive to such things as solvent composition and temperature in the environment of the helices, and a_0 presumably includes all the (nonhelical) background rotations. Since the ORD of the disordered form of polypeptides and proteins obeys the Moffitt equation as well as the simple Drude equation, we can, following Urnes' suggestion (1963), rewrite the Moffitt equation as

$$[m'] = (f_\alpha a_0^\alpha + f_c a_0^c)/(\lambda^2 - \lambda_0^2) + (f_\alpha b_0^\alpha + f_c b_0^c)/(\lambda^2 - \lambda_0^2)^2$$

or

$$[m'] = [f_\alpha(a_0^\alpha - a_0^c) + a_0^c]/(\lambda^2 - \lambda_0^2) + [f_\alpha(b_0^\alpha - b_0^c) + b_0^c]/(\lambda^2 - \lambda_0^2)^2. \quad (58)$$

Here we assume that $(a_0^\alpha - a_0^c)$ and $(b_0^\alpha - b_0^c)$ cancel the residue rotations in the helical and coiled forms and represent the a_0 and b_0 values of the helical backcone only. (For a somewhat different formulation but the same final conclusion, see Yang, 1967a.) From Eq. (58) we have

$$\text{Fraction of helix,} \, f_\alpha = (b_0 - b_0^c)/(b_0^\alpha - b_0^c) \quad (59)$$

which reduces to Eq. (52) when b_0^c is assumed to be zero.

Equation (58) introduces a nonhelical b_0^c, which at present is not exactly known. For example, coiled poly-L-glutamic acid in water has a λ_c of about 204 mμ and when cast into the Moffitt equation it gives a b_0^c of about $+70$ (Iizuka and Yang, 1965). But this synthetic polypeptide does not resemble the amino acid residues in a protein molecule, which can be very rigid and compact. All we know is that the λ_c of coiled poly-L-glutamic acid does not vary much under various conditions, and the magnitude of its k in Eq. (38) is reduced when the molecule is contracted. Thus, the b_0^c in Eq. (59) for a protein molecule might be expected to be even smaller than $+70$ for poly-L-glutamic acid coils and perhaps close to zero. Neglect of this term may not seriously affect the estimated helical content, although the uncertainty remains and the estimates could easily be off by 5–10% helical content.

Second, the basic premise for using synthetic polypeptides as reference compounds is that the structures of proteins and polypeptides are similar. The calculations for the proteins, however, are really the *effective* helical contents which would have been observed with hypothetical mixtures of appropriate amounts of helices and coils of a model compound. In a globular protein molecule the helical segments are usually short, unlike the high-molecular-weight helical polypeptides. Both theory (Tinoco *et al.*, 1963) and experiments (Goodman and Rosen, 1964) show that the magnitude of b_0 and the 233-mμ trough of short helices increase with the degree of polymerization and reach a plateau above a certain critical chain length. Thus, the use of $b_0^\alpha \cong -630$ for proteins having short helices would underestimate the helical contents. On the other hand, the presence of a number of residues in the disordered region of a protein molecule with angles appropriate for the helices may contribute to the b_0^α, even though the residues are not part of the helical segments. The difficulty lies in the definition of a residue as helical. The contributions of these residues could overestimate the helical content. Obviously some partial cancellation is expected from these opposite effects. The resultant effects might be only secondary to the net b_0 observed.

Third, the empirical parameter, $b_0^\alpha \cong -630$, is now being reinvestigated under various conditions, since the Moffitt theory can no longer predict a constancy of b_0^α with respect to solvent composition and temperature. The magnitude of b_0 was found to decrease gradually with increasing temperature; for example, the b_0 of poly-γ-benzyl-L-glutamate in dimethyl-formamide and ethylene dichloride was about -580 to -600 at 60°C instead of -630 at room temperature (Sarkar and Yang, unpublished data). Franzen *et al.* (1967) reported a further change of b_0 to about -500 for the same polymer in N-methylcaprolactam at 90–150°C. Whether any small change in conformation accompanies such drop in $-b_0$ is still not easy to detect. The effect of solvent composition has been studied by Cassim and Taylor (1965) for poly-γ-benzyl-L-glutamate; the b_0^α was found to be a linear function of the refractive index of the solvent (measured at sodium D line)

$$b_0^\alpha = 730.3 n_D - 1701 \tag{60}$$

with one notable exception. The helix-promoting solvents, except m-cresol, shown in Fig. 10, all have a refractive index of about 1.4. Ironically, b_0^α in m-cresol is also close to the average value of -630 found in the middle 1950's instead of -576 based on Eq. (60), which had led to the conclusion that b_0^α was independent of the solvent used. Equation (60) is based on measurements of one polypeptide; its general applicability to other helical polypeptides remains to be proved. It is also not known whether variations of the background rotations with solvent may lead to different nonhelical b_0 in Eq. (58), thus accounting partially for the linear relation of Eq. (60) (Yang, 1967a).

The effect of refractive index of the solvent on b_0 does not present a problem if we are only interested in aqueous solutions. There is no *a priori* reason that Eq. (60) for organic solvents must be equally applicable to aqueous solutions. On the other hand, it is always possible that water-soluble helical polypeptides should have a b^α of -730 based on Eq. (60) instead of -630 observed experimentally, and the lower magnitude could have indicated an imperfect helix.

Fourth, the Cotton effects produced by nonpeptide chromophores present another problem. Obviously, the Moffitt equation will no longer be applicable in the usual wavelength range of 300–600 mμ if the optically active bands also appear in the same region, e.g. the Soret bands of heme proteins. One must then treat the data in regions away from the absorption bands, as Urnes (1964, 1965) did on myoglobin. The Cotton effects of aromatic groups and disulfide bonds below 300 mμ will also affect the rotations in the visible region. Although sporadic experiments seem to show a small effect on the b_0, the data are just beginning to accumulate and it is too early to draw any conclusions. While nonpeptide optical activity is a problem in one sense, it provides a new means of studying the tertiary structure of a protein molecule involving these side groups.

Question has also been raised as to the proper choice of λ_0 in the Moffitt equation for proteins. In a helical polypeptide the rotatory contributions of the backbone dominate, and computer analyses indicate that $\lambda_0 = 212$ mμ gives the best fit. A different λ_0 may be found for Eq. (58) when the helical content is low (Sogami *et al.*, 1963). Recent calculations of the Moffitt parameters by the least-square method further suggest that such evaluation becomes indeterminate as b_0 approaches zero (DeTar, 1966). The adoption of a common λ_0 has the advantage of putting all ORD data on the same basis and enabling us to compare the relative helical contents among various proteins.

Most of what we have discussed can of course be equally applied to the trough or peak method and the CD methods. In view of the many problems in the phenomenological treatments, it is indeed surprising that the Moffitt equation so far stands well. In the middle 1950's ORD suggested a helical content of about 30% in lysozyme, which is not too different from the 35% based on recent X-ray studies (Blake *et al.*, 1965). The estimated helical content of myoglobin is between 70 and 80% (see, for example, Urnes, 1964, 1965), again in good agreement with X-ray results. As more three-dimensional structures of proteins are unveiled, the reliability of the ORD and CD methods will be subject to a meaningful test.

(d) Coexistence of the Helix and β-form

The β-form introduces an additional complication in the ORD analyses. The Moffitt equation assumes that the protein in the ORD has only one kind of secondary structure, the α-helix; the b_0 method obviously is not sufficient to solve two unknowns. Imahori (1960) first found that the ORD of the β-form also obeys the Moffitt equation (see also Wada et al., 1961). Thus, in principle we can utilize the a_0 that is also conformation-dependent and solve two simultaneous equations (cf. Eq. (50):

$$a_0 = f_\alpha a_0^\alpha + f_\beta a_0^\beta + (1 - f_\alpha - f_\beta) a_0^c \tag{61a}$$

and

$$b_0 = f_\alpha b_0^\alpha + f_\beta b_0^\beta + (1 - f_\alpha - f_\beta) b_0^c \tag{61b}$$

provided that all the reference values are known. But the separation of a_0 into three components is arbitrary and uncertain, and the reference values of a_0^β and b_0^β are still very much in doubt (for instance, b_0^β varies between ± 200 according to the few data reported in the literature (Yang, 1967a)). Therefore attempts to calculate the percentages of the helical and β-forms of proteins such as lysozyme and β-lactoglobulin at this stage are risky at best and perhaps premature.

Since the b_0^β appears to be small as compared with the b_0^α, we can still estimate the helical content as if the protein molecule contains no β-form. The possible errors that will result in such calculation can be determined from the following equation (Yang, 1967b):

$$f_\alpha = (b_0^c - b_0)/(b_0^c - b_0^\alpha) + f_\beta (b_0^\beta - b_0^c)/(b_0^c - b_0^\alpha) \tag{62a}$$

or

$$f_\alpha(\text{app}) = f_\alpha + f_\beta (b_0^c - b_0^\beta)/(b_0^c - b_0^\alpha). \tag{62b}$$

A zero b_0^β and b_0^c would of course not affect the estimate of the α-helix. If b_0^β is positive (or negative) and $b_0^c \cong 0$, the apparent f_α will be less (or more) than the true f_α. Taking the limits of $b_0^\beta = \pm 200$, we find that every 3 or 4% β-form will reduce or raise the estimated helical content by about 1%, depending on the sign of b_0^β. The helical form would usually dominate the experimental b_0 when both forms coexist in a protein molecule, and the current b_0 method still permits a reasonable estimate of the helical content, even if the molecule contains a moderate amount of the β-form.

Similarly, the amount of the helical and β-forms can in principle be determined with simultaneous measurements of the 233-$m\mu$ trough and 198-$m\mu$ peak. Here again, the reference values are uncertain; besides, the possible rotations due to the presence of nonpeptide chromophores are not known. Just as in Eq. (62), it can easily be shown (Iizuka and Yang, 1966) that at 233 $m\mu$:

$$f_\alpha = [([m'] - [m']^c) - f_\beta([m']^\beta - [m']^c)]/([m']^\alpha - [m']^c) \tag{63a}$$

or

$$f_\alpha(\text{app}) = f_\alpha + f_\beta([m']^\beta - [m']^c)/([m']^\alpha - [m']^c). \tag{63b}$$

Assuming $[m']^\alpha \cong -15,000$, $[m']^\beta \cong -6000$, and $[m']^c \cong -2000$ (Table I), every 3 or 4% β-form, when present in a protein molecule, would raise the estimated helical content by about 1% (neglecting the rotations of nonpeptide chromophores).

The best procedure for the ORD and CD analyses of proteins seems to be to measure both of them throughout the entire attainable wavelength range (from 600 to 185 mμ with current spectropolarimeters and circular dichrometers). Direct observation of a 233-mμ trough, a 198-mμ peak, and a double CD minimum at 222 and 210 mμ would be a strong indication of the presence of the α-helix. Their magnitude also provides a rough estimate of the helical content. The more satisfactory estimate from the b_0 method of the Moffitt equation also becomes more meaningful. A protein molecule containing only the β-form presents no problem in its identification. Our present knowledge, however, does not warrant any quantitative analysis of the β-form, if both the helix and β-form coexist in a protein molecule.

8.   CONFORMATIONAL TRANSITIONS

Almost any physical property that characterizes the protein conformation can be utilized to monitor the conformational transitions of the protein molecule. The choice depends upon the sensitivity of the physical method, the quantitative analysis of the measurements and the ease of the operation. Optical rotation has long been used in studying the protein denaturation; it has become increasingly popular because of our current understanding of the conformational changes that lead to the changes in rotations. Among other physical techniques that have also been employed for such studies are viscometry, infrared and ultraviolet spectroscopy, and deuterium-hydrogen exchange.

(a) Helix–Coil Transition

Of various conformational changes the helix–coil transition has been most intensively studied. Poly-γ-benzyl-L-glutamate was the first synthetic polypeptide subject to detailed studies by optical rotation, viscometry and light scattering. Figure 10 has already shown that the helical conformation is stable in some poor solvents and the coiled form in other good solvents. Thus we can look for a transition between the two conformations by systematically varying composition of the mixed solvents of the two classes. Such is the case in Fig. 20, where the rotations reveal a very sharp, reversible transition in 1:4 (w/w) ethylene dichloride–dichloroacetic acid at room temperature. Correspondingly, the intrinsic

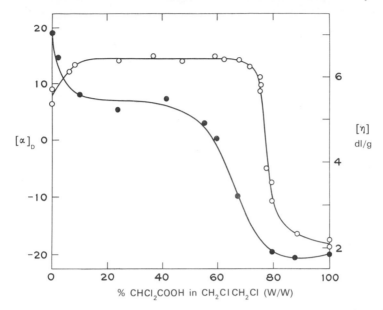

FIG. 20. Specific rotation (at 589 mμ) and intrinsic viscosity of poly-γ-benzyl-L-glutamate as a function of solvent composition. (Yang and Doty, 1956.)

viscosity, which monitors the shape and hydrodynamic volume of the molecules, shows a gradual fall in the transition region and reaches a value characteristic of the coiled form at about 80% (w/w) dichloroacetic acid. Imahori and Doty, as quoted by Doty (1957), attributed this to the formation of some intermolecular β-form (based on infrared evidence), when the first breaks occurred in the helix with increasing concentration of dichloroacetic acid. This, in turn, caused a contraction and aggregation of the polymer that affected the viscosity out of proportion to the number of residues that have actually been removed from the helical conformation. The tiny region of β-form, however, appeared on the helix side of the transition and disappeared when the coiled form became predominant. (There was also an initial fall in intrinsic viscosity and rise in rotation when dichloroacetic acid was first added to ethylene dichloride. This was probably related to the initial solvation of the polypeptide chain by dichloroacetic acid, which reduced the axial ratio of the rod and changed its optical properties.) Similar sharp transitions were also observed in other pairs of solvents and could be induced by adding a small amount of a nonsolvent such as water to a good solvent such as dichloroacetic acid well before the precipitation of the polypeptide (Yang and Doty, 1956).

Instead of varying the solvent composition, the conformation of poly-γ-benzyl-L-glutamate at the transition region can be changed with temperature; for instance, in 80% dichloroacetic acid (Fig. 20) a change from room temperature to 40°C converted the coiled form to the helical conformation. Unlike protein denaturation, which favors the disordered form at high temperature, the "reverse" transition in this case can be explained purely on a thermodynamic basis (Doty and Yang, 1956). The subject of the helix–coil transition has since stimulated many theoretical studies, which provide a quantitative interpretation of the observed phenomenon (Gibbs and DiMarzio, 1959; Peller, 1959; Zimm and Bragg, 1959; Lifson and Roig, 1961).

Since the stability of the helix is a cooperative effect of the hydrogen bonding, the sharpness of the transition is therefore dependent on the molecular weight and molecular weight distribution of the polypeptide. Poly-γ-benzyl-L-glutamate (Doty and Yang, 1956; Zimm et al., 1959) is one example. In Section 5 (a) we mentioned that the L-polypeptides usually favor the right-handed helices, and that small fractions of D-residues randomly incorporated into the L-polypeptides would retain the same right-handed sense of screw. But the inclusion of D-residues, which normally favor the left-handed helix, will weaken the helical conformation of the L-polypeptides and thereby shift the transition region toward lower percentage of dichloroacetic acid in mixed solvents (Fig. 21).

In water-soluble polypeptides containing ionizable groups, such as poly-L-glutamic acid, the helices are stabilized against the electrostatic repulsion of the carboxyl groups in the side chain. Figure 22 shows that the helix–coil transition can be brought about merely by adjusting the pH of the solution. The corresponding intrinsic viscosity in this case indicated a sharp minimum near pH 6, which suggests that the partial helices had smaller hydrodynamic volume than the coiled form (above pH 7). Note that the viscosity of a typical polyelectrolyte anion would be expected to fall monotonically with pH, if no transition occurred. Figure 22 also shows that the helix–coil transition of poly-L-glutamic acid takes place almost entirely within the region of 35–80% ionization of the carboxyl groups in the side chain. More recently, the addition of a nonaqueous solvent such as dioxane (more than 50%) to the aqueous solution of poly-L-glutamic acid was reported to induce the transition even at neutral pH (apparent), because the pK of the carboxylate groups increases markedly in the presence of, say, 50% dioxane (Fig. 23).

(b) Coiled-β and Helix-β Transitions

In Section 5(a) we mentioned that the oligopeptides of γ-benzyl-L-glutamate form β-aggregates in concentrated solutions, which dissociate into the coiled form upon dilution. The intermolecular β-form of

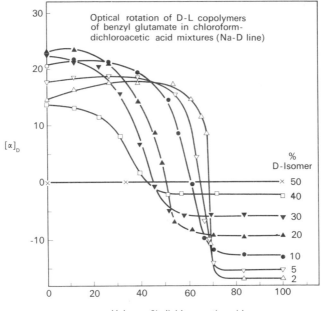

FIG. 21. Specific rotation (at 589 mμ) of D-L-copolymers of γ-benzyl-glutamate in chloroform–dichloroacetic acid mixtures. The lines on the right indicate the calculated rotations assuming that the rotations of D- and L-residues are additive. The curve for the pure L-polypeptide (not shown) falls close to the copolymer containing 2% D-isomers. (Blout *et al.*, 1957.)

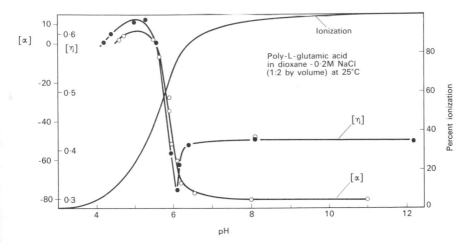

FIG. 22. The pH dependence of specific rotation (at 589 mμ), intrinsic viscosity, and degree of ionization for poly-L-glutamic acid in 2 : 1 (v/v) 0.2 M NaCl-dioxane. (Doty *et al.*, 1957.)

polymers, however, are usually insoluble in any solvent unless the secondary structure is broken up. Poly-O-acetyl-L-serine (Fasman and Blout, 1960) and poly-O-benzyl-L-serine (Bradbury *et al.*, 1962), for instance, can be dissolved in a good solvent such as dichloroacetic acid. Upon dilution with a poor solvent such as chloroform, the polymers undergo a sharp coil-β transition (based on evidence of infrared spectra). An example is illustrated in Fig. 24. Both polymers are believed to contain partial cross-β form in the mixed solvent. On the other hand, the β-form of poly-O-acetyl-L-serine gels on storing if the solution contains

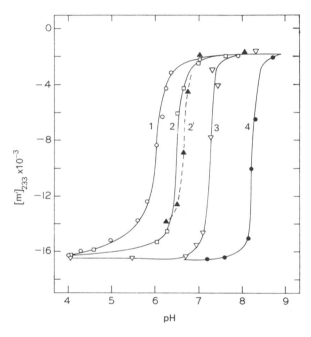

FIG. 23. Helix–coil transition of poly-L-glutamic acid in dioxane–water solutions. Percent dioxane (v/v): 1, zero, 2, 10; 2′, 10 with 0.2 M KF; 3, 30; and 4, 50. (Iizuka and Yang, 1965.)

less than 30% dichloroacetic acid. This gelation strongly depends on the polymer concentration, suggesting the presence of intermolecular aggregation of the cross-β form.

The conversion of the disordered form of silk fibroin into the cross-β form when dioxane or methanol was added to the aqueous solution has been discussed (Section 5 (b)). Sarkar and Doty (1966) reported that the coiled form of poly-L-lysine at neutral pH also underwent a coil-to-β transition when a small amount of sodium dodecyl sulfate was added to a dilute solution of the polymer. This transition was absent for mixtures of poly-L-glutamic acid and a cationic detergent. The mechanism of this transition is still not clear.

Solutions of helical polypeptides often develop a transient haze during helix–coil transition in mixed solvents when the solvent composition is slowly adjusted; this suggests a complex equilibrium among the helix, coil and β-form in the intermediate state. Poly-L-lysine in water exists in the helical form at or above pH 11, which on mild heating at 50°C undergoes an α-to-β transition (Sarkar and Doty, 1966; Davidson *et al.*, 1966; Townend *et al.*, 1966). Another interesting example is that

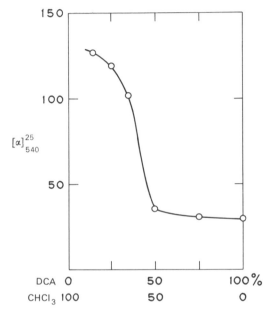

FIG. 24. Specific rotation (at 589 mμ) of poly-O-acetyl-L-serine in chloroform–dichloroacetic acid mixtures. (Fasman and Blout, 1960.)

the helical polypeptides such as poly-L-glutamic acid and poly-L-lysine in solid state can be induced to form the β-form merely by controlling the relative humidity, low humidity favoring the β-form (Elliott *et al.*, 1957; Blout and Lenormant, 1957; Lenormant *et al.*, 1958). The X-ray evidence for humidity-controlled conformational transitions in poly-L-lysine has now been reported by Shmueli and Traub (1965).

The conformational transitions are much simpler to interpret in model polypeptides than in proteins. Change in rotations of proteins does not necessarily indicate that only one type of transition being monitored is occurring. If measurements can be made in the ultraviolet

region below 250 mμ, it is desirable to choose a wavelength characteristic of certain conformation, for instance, the 233-mμ trough for the
helix. The reduction in $-[m']_{233}$ and $[m']_{198}$ would be a good indication
of the disruption of the helix. The shift of the trough to 230 mμ and the
peak to above 200 mμ, together with a decrease in the trough and peak
magnitude, would suggest an α-to-β transition. On the other hand, a
helix–coil transition would lead to a Cotton effect profile resembling that
of the coiled form (Figs. 12 and 13). The same argument can be applied
to the β-coil transition, although few data are available on this subject.

9. CONCLUSIONS

The observations of the conformational rotations of helical polypeptides in the middle 1950's began a rapid development of ORD for
studying protein conformations. Presently the Moffitt equation applicable
for the visible rotatory dispersion of proteins provides a reasonable
estimate of helical content in a protein molecule. In the early 1960's
the recording spectropolarimeters extended ORD measurements to
about 185 mμ, and enabled us to observe the Cotton effects of the peptide
chromophores. The complementary CD measurements also became
increasingly popular with the introduction of commercial circular
dichrometers. Together ORD and CD provide a wealth of information
and firmly establish the origin of optical activity of protein conformations.
These new developments also help us to use the Moffitt equation with
more confidence. Since most organic solvents absorb strongly in the
ultraviolet region, measurements of the Cotton effects of proteins are
largely confined to aqueous solutions or a few other solvents such as
methanol and dioxane.

The working hypothesis that proteins are made up of partial helices
and nonhelical segments, albeit oversimplified, has been used to great
advantage in our probing of the secondary structure of proteins. This
idea has since been adopted for many other physical methods. The
recent discovery of the β-form does complicate the phenomenological
treatments, but the ORD and CD of the β-form are overshadowed by
those of the helical form when they coexist in a protein molecule. Thus
moderate amount of the β-form, if present, would not introduce significant errors in the estimated helical content. Among the problems that
remain to be solved are: the more accurate reference values of various
conformations, the complications due to the Cotton effects of nonpeptide
chromophores, and the ORD of tertiary and quaternary structures of
proteins. Any empirical treatment permits only cautious interpretation
of experimental results. On the other hand, few physical methods that
are currently used can provide as much information as ORD and CD

about conformations and, in particular, conformational changes of proteins in solution. It is almost certain that these two techniques will play an even more important role in advancing our knowledge of protein conformations for years to come.

ACKNOWLEDGEMENTS

The author is indebted to Professor George M. Holzwarth for his contribution on the section on instrumentation. He also thanks Mrs. F. Kasarda for her assistance during the preparation of the manuscript.

This work was aided by the U.S. Public Health Service grants (GM-K3-3441, GM-10880, HE-06285).

APPENDIX

TABLE A1. NUMERICAL CONSTANTS FOR CALCULATING b_0 OF THE MOFFITT EQUATION ($\lambda_0 = 212$ mμ) *

λ, mμ	λ^2, μ^2	$(\lambda^2/\lambda_0^2 - 1)$	$1/(\lambda^2/\lambda_0^2 - 1)$	$3/(n^2 + 2)$ for water at 20°†
600	0.3600	7.010	0.1427	0.7945
580	0.3364	6.485	0.1542	0.7941
560	0.3136	5.978	0.1673	0.7938
540	0.2916	5.488	0.1822	0.7933
520	0.2704	5.016	0.1993	0.7929
500	0.2500	4.563	0.2192	0.7924
480	0.2304	4.126	0.2423	0.7918
460	0.2116	3.708	0.2697	0.7912
440	0.1936	3.308	0.3023	0.7904
420	0.1764	2.925	0.3419	0.7896
400	0.1600	2.560	0.3906	0.7886
380	0.1444	2.213	0.4519	0.7875
360	0.1296	1.884	0.5309	0.7861
340	0.1156	1.572	0.6361	0.7845
320	0.1024	1.278	0.7822	0.7825
300	0.0900	1.003	0.9975	0.7799
290	0.0841	0.8712	1.148	0.7784
280	0.0784	0.7444	1.343	0.7767
270	0.0729	0.6220	1.608	0.7748
260	0.0676	0.5041	1.984	0.7725
250	—	—	—	0.7698
240	—	—	—	0.7667
233	—	—	—	0.7641
230	—	—	—	0.7629
220	—	—	—	0.7584
210	—	—	—	0.7527
205	—	—	—	0.7494
200	—	—	—	0.7456
198	—	—	—	0.7440
190	—	—	—	0.7364
185	—	—	—	0.7306

*Taken from Yang (1967a).

†The refractive indices of water were calculated from the Duclaux–Jeantet formula: $n^2 = 1.762530 - 0.0133998\lambda^2 + 0.00630957/(\lambda^2 - 0.0158800)$, $T = 20°C$ (λ in microns) (Dorsey, 1940).

TABLE AII. THE LORENTZ CORRECTION, $3/(n^2+2)$, OF SEVERAL ORGANIC SOLVENTS*

λ, mμ	Chloroform	Chlorohydrin	Dichloro-acetic acid	Dimethyl-formamide	p-Dioxane	Ethylene dichloride	8M Urea (aq.)
600	0.7351	0.7364	0.7238	0.7433	0.7477	0.7352	0.7578
580	0.7346	0.7360	0.7233	0.7427	0.7473	0.7348	0.7573
560	0.7340	0.7355	0.7227	0.7420	0.7468	0.7343	0.7568
540	0.7334	0.7350	0.7221	0.7412	0.7463	0.7337	0.7563
520	0.7327	0.7344	0.7214	0.7404	0.7457	0.7331	0.7557
500	0.7320	0.7337	0.7206	0.7394	0.7451	0.7324	0.7550
480	0.7311	0.7330	0.7198	0.7383	0.7444	0.7315	0.7542
460	0.7301	0.7321	0.7187	0.7370	0.7435	0.7306	0.7532
440	0.7289	0.7311	0.7176	0.7356	0.7426	0.7296	0.7522
420	0.7276	0.7297	0.7162	0.7339	0.7415	0.7283	0.7510
400	0.7261	0.7286	0.7147	0.7319	0.7402	0.7269	0.7496
380	0.7242	0.7270	0.7128	0.7295	0.7387	0.7252	0.7479
360	0.7221	0.7251	0.7106	0.7268	0.7369	0.7232	0.7459
340	0.7194	0.7229	0.7080	0.7234	0.7347	0.7209	0.7435
320	0.7163	0.7202	0.7048	0.7192	0.7321	0.7180	0.7407
300	0.7123	0.7168	0.7009	0.7141	0.7289	0.7144	0.7371
290	0.7100	0.7148	0.6985	0.7110	0.7270	0.7122	0.7349
280	0.7074	0.7126	0.6959	0.7075	0.7248	0.7099	0.7326
270	0.7044	0.7100	0.6929	0.7035	0.7224	0.7071	0.7298
260	0.7010	0.7072	0.6895	0.6989	0.7196	0.7040	0.7267
250	0.6971	0.7039	0.6856	0.6936	0.7165	0.7005	0.7232

*Based on data of Foss and Schellman (1964).

87

TABLE AIII. NUMERICAL VALUES OF
THE FUNCTION $e^{-c^2} \int_0^c e^{x^2} dx$ *†‡

c	$e^{-c^2} \int_0^c e^{x^2}\, dx$	c	$e^{-c^2} \int_0^c e^{x^2}\, dx$
0	0.00000	1.2	0.50726
0.1	0.09903	1.3	0.48339
0.2	0.19465	1.4	0.4565
0.3	0.27620	1.5	0.4282
0.4	0.35943	1.6	0.4000
0.5	0.42444	1.8	0.3468
0.6	0.47477	2.0	0.30135
0.7	0.51050	2.2	0.2629
0.8	0.53210	2.4	0.2353
0.9	0.54073	2.6	0.2122
0.93	0.54224	2.8	0.1936
1.0	0.53806	3.1623	0.1667
1.1	0.52620	>3.2	$1/(2c)$

*Taken from Lowry (1935).
†For negative c, values of the function are negative.
‡The function can also be evaluated from the convergent series:

$$e^{-c^2} \int_0^c e^{x^2}\, dx$$

$$= \sum_{n=1}^{\infty} (-1)^{n-1} \cdot 2^{2(n-1)} \cdot (n-1)!\, c^{2n-1}/(2n-1)!$$

$$= c - 2c^3/(1 \cdot 3) + 2^2 c^5/(1 \cdot 3 \cdot 5) - 2^3 c^7/(1 \cdot 3 \cdot 5 \cdot 7)$$
$$+ 2^4 c^9/(1 \cdot 3 \cdot 5 \cdot 7 \cdot 9) \ldots + \ldots$$

NOTE ADDED IN PROOF

J. Thiery and I. Tinoco, Jr. (private communication) have developed a computer program for CD calibration. For a compound having a well-defined CD band, we can correlate the experimental rotations with those calculated from the CD (using Eq. (16)) by the relationship

$$[m]_{\text{exptl}} = f[m]_{\text{calc}} + a_c \lambda_c^2/(\lambda^2 - \lambda_c^2). \qquad (64)$$

Here the Drude term represents the sum of rotations of all other CD bands distant from the absorption band under consideration (background rotations) (see Eqs. (37) and (38)). The factor f corrects any imperfect calibration of the CD instrument. J. Y. Cassim of this laboratory tested the Thiery-Tinoco program with several compounds and found perfect agreement between the experimental and computed rotations with a consistent f value. The CD instrument was then adjusted so that f became unity. Our $(\epsilon_L - \epsilon_R)$ for d-10-camphorsulfonic acid in water was $+2.20$ at 291 mμ and the corresponding rotation at the peak (306 mμ) was $+4.120$. For d-camphor in dioxane, $(\epsilon_L - \epsilon_R)$ was $+1.69$ at 300 mμ, as compared with $+1.6$ reported by Velluz *et al.* (1965). Note that the absolute magnitude of these values may depend on the purity of the samples, but they will not affect the factor f for a particular instrument, since both ORD and CD are measured on the same samples.

Having overcome the calibration problem, Cassim (manuscripts in preparation) repeated the computer calculations shown in Figs. 12 and 13. There was a perfect agreement between experimental and computed rotations for the α-helix, suggesting that the CD bands below 180 mμ do not contribute significantly above 180 mμ (although Moffitt had predicted another band splitting near 150 mμ) and, further, the rotations in the visible region (as in Fig. 11) arise from the three observed CD bands. The computed visible rotatory dispersion is much like the experimental profile, but the magnitude and even the sign of the calculated values could change, depending on the extrapolation of the CD spectrum near the instrumental limits and/or experimental errors. Note that the visible rotations are extremely small and are the result of the sum of large positive and negative rotations due to the three CD bands. Nevertheless, the b_0, not a_0, of the Moffit equation does not change much with small deviations in the extrapolation of the CD bands. (For a different method of analysis, see Carver *et al.* (1966a).) Agreement between experimental and computed rotations was equally good for the coiled form (after proper CD calibration). The agreement was far better than that shown in Fig. 13, except for some discrepancies regarding the small Cotton effects near 240 mμ. Comparison of the experimental and computed rotations for the β-form of poly-L-lysine, however, suggests there may be other CD bands below 180 mμ that also contribute rotations above 180 mμ.

The ORD and CD of the β-forms have now received wide attention. Infrared studies suggest that both silk fibroin and poly-L-lysine (Section 5 (b) and (c)) form the anti-parallel β-form. Recently, E. Iizuka (private communication) studied the oriented films of silk fibroin cast from 50% methanol and water, and concluded that the β-form of the former is intramolecular, in contrast to the intermolecular type of the latter. The question of intra- *versus* inter-molecular forms remains unsolved for the β-form of poly-L-lysine. Note that the magnitude of rotations can change to some extent if there is extensive aggregation; this seems to be true of the intramolecular β-form of silk fibroin (Iizuka and Yang, to be published).

REFERENCES

ADKINS B. J. and YANG J. T. (1967) *Biochemistry*, in press.

ANUFRIEVA E. V., VOLCHEK B. Z., ILLARIONOVA N. G., KALIKHEVICH V. N., KOROTKINA O. Z., MITIN YU. V., PTITSYN O. B., PURKINA A. V. and ESKIN V. E. (1965) *Biofizika* **10**, 346.

BALASUBRAMANIAN D. and WETLAUFER D. B. (1967) In G. N. RAMACHANDRAN (ed.), *Conformation of Biopolymers*, Academic Press, New York, p. 147.
BEYCHOK S. (1966) *Science*, **154**, 1288.
BEYCHOK S. (1967) In G. D. FASMAN (ed.), *Poly-α-Amino Acids*, Marcel Dekker, New York, chap. 7.
BEYCHOK S. and BLOUT E. R. (1961) *J. Mol. Biol.* **3**, 769.
BEYCHOK S. and FASMAN G. D. (1964) *Biochemistry* **3**, 1675.
BILLINGS B. H. (1949) *J. Opt. Soc. Am.* **39**, 797, 802.
BLAKE C. C. F., KOENIG D. F., MAIR G. A., NORTH A. C. T., PHILLIPS D. C. and SARMA V. R. (1965) *Nature* **206**, 757.
BLOUT E. R. and LENORMANT H. (1957) *Nature* **179**, 960.
BLOUT E. R., DOTY P. and Yang J. T. (1957) *J. Am. Chem. Soc.* **79**, 749.
BRADBURY E. M., BROWN L., DOWNIE A. R., ELLIOTT A., FRASER R. D. B. and HANBY W. E. (1962) *J. Mol. Biol.* **5**, 230.
BRADLEY D. F., TINOCO I., Jr. and WOODY R. W. (1963) *Biopolymers* **1**, 239.
BRESLOW E., BEYCHOK S., HARDMAN K. D. AND GURD F. R. N. (1965) *J. Biol. Chem.* **240**, 304.
CARVER J. P., SHECHTER E. and BLOUT E. R. (1966a) *J. Am. Chem. Soc.* **88**, 2550.
CARVER J. P., SHECHTER E. and BLOUT E. R. (1966b) *J. Am. Chem. Soc.* **88**, 2562.
CARY H., HAWES R. C., HOOPER P. B., DUFFIELD J. J. and GEORGE K. P. (1964) *Applied Optics* **3**, 329.
CASSIM J. Y. and TAYLOR E. W. (1965) *Biophys. J.* **5**, 553.
CASSIM J. Y. and YANG J. T. (1966) *Biochem. Biophys. Res. Commun.* **26**, 58.
CHIGNELL D. A. and GRATZER W. B. (1966) *Nature* **210**, 262.
CRICK F. H. C. and KENDREW J. C. (1957) *Advances in Protein Chemistry* **12**, 134.
DAVIDSON B., TOONEY N. and FASMAN G. D. (1966) *Biochem. Biophys. Res. Commun.* **23**, 156.
DETAR D. F. (1966) *Biophys. J.* **6**, 505.
DORSEY N. E. (1940) *Properties of Ordinary Water-Substances*, Reinhold, New York, p. 280.
DOTY P. (1957) *Collection Czechoslov. Chem. Commun.* **22**, 5.
DOTY P. and YANG J. T. (1956) *J. Am. Chem. Soc.* **78**, 498.
DOTY P., HOLTZER A. M., BRADBURY J. H. and BLOUT E. R. (1954) *J. Am. Chem. Soc.* **76**, 4493.
DOTY P., WADA A., YANG. J. T. and BLOUT E. R. (1957) *J. Polymer Sci.* **23**, 851.
ELLIOTT A., MALCOLM B. R. and HANBY W. E. (1957) *Nature* **179**, 960.
FASMAN G. D. and BLOUT E. R. (1960) *J. Am. Chem. Soc.* **82**, 2262.
FASMAN G. D., BODENHEIMER E. and LINDBLOW C. (1964) *Biochemistry* **3**, 1665.
FASMAN G. D., LANDSBERG M. and BUCHWALD M. (1965) *Can. J. Chem.* **43**, 1588.
FOSS J. G. and SCHELLMAN J. A. (1964) *J. Chem. Eng. Data* **9**, 55.
FRANZEN J. S., HARRY J. B. and BOBIK C. (1967) *Biopolymers* **5**, 93.
FRIED D. L. (1965) *Applied Optics* **4**, 79.
GIBBS J. H. and DiMARZIO E. A. (1959) *J. Chem. Phys.* **30**, 271.
GILLAM A. E. and STERN E. S. (1957) *An Introduction to Electronic Absorption Spectroscopy in Organic Chemistry*, Edward Arnold, London.
GILLHAM E. J. and KING R. J. (1961) *J. Sci. Instr.* **38**, 21.
GLAZER A. N. and SIMMONS N. S. (1966) *J. Am. Chem. Soc.* **88**, 2335.
GOODMAN M. and ROSEN I. G. (1964) *Biopolymers* **2**, 537.
GRATZER W. B., HOLZWARTH G. M. and DOTY P. (1961) *Proc. Natl. Acad. Sci. U.S.* **47**, 1785.
GRAU G. K. (1965) *Applied Optics* **4**, 755.
GROSJEAN M. and LEGRAND M. (1960) *Compt. Rend.* **251**, 2150.
GROSJEAN M. and TARI M. (1964) *Compt. Rend.* **258**, 2034.
HARRAP B. S. and STAPLETON I. W. (1963) *Biochim. Biophys. Acta* **75**, 31.
HARRINGTON W. F. and VON HIPPEL P. H. (1961) *Advances in Protein Chemistry* **16**, 1.
HARRINGTON W. F., JOSEPHS R. and SEGAL D. M. (1966) *Ann. Rev. Biochem.* **35**, 599.
HARRIS T. L., HIRST E. L. and WOOD C. E. (1932) *J. Chem. Soc. (London)*, **2**108.

HARRISON S. C. and BLOUT E. R. (1965) *J. Biol. Chem.* **240**, 299.
HOLZWARTH G. M. (1964) Ph. D. Dissertation, Harvard University.
HOLZWARTH G. M. (1965)*Rev. Sci. Instr.* **36**, 59.
HOLZWARTH G. M. and DOTY P. (1965)*J. Am. Chem. Soc.* **87**, 218.
IIZUKA E. and YANG J. T. (1965) *Biochemistry* **4**, 1249.
IIZUKA E. and YANG J. T. (1966) *Proc. Natl. Acad. Sci. U.S.* **55**, 1175.
IKEDA S., MAEDA H. and ISEMURA T. (1964)*J. Mol. Biol.* **10**, 223.
IMAHORI K. (1960) *Biochim. et Biophys. Acta* **37**, 336.
IMAHORI K. (1963) *Kobunshi (Japan)* **12**, Suppl. 1, 34.
IMAHORI K. and YAHARA I. (1964) *Biopolymers Symposia* **1**, 421.
JIRGENSONS B. (1961) *Tetrahedron* **13**, 166.
KAMASHIMA K. (1966)*J. Phys. Soc. Japan* **21**, 1781.
KAUZMANN W. (1957) *Quantum Chemistry*, Academic Press, New York, chap. 15.
KENDREW J. C., WATSON H. C., STRANDBERG B. E., DICKERSON R. E., PHILLIPS D. C. and SHORE V. C. (1961) *Nature* **190**, 666.
LENORMANT H., BANDRAS A. and BLOUT E. R. (1958)*J. Am. Chem. Soc.* **80**, 6191.
LIFSON S. and ROIG A. (1961)*J. Chem. Phys.* **34**, 1963.
LINDERSTRØM-LANG K. and SCHELLMAN J. A. (1954) *Biochim. Biophys. Acta* **15**, 156.
LITMAN B. J. and SCHELLMAN J. A. (1965)*J. Phys. Chem.* **69**, 978.
LOWRY T. M. (1935) *Optical Rotatory Power*, Longmans, Green, London; Dover Publications, 1964.
MITCHELL S. (1957)*J. Sci. Instr.* **34**, 89.
MOFFITT W. (1956)*J. Chem. Phys.* **25**, 467.
MOFFITT W. and MOSCOWITZ A. (1959)*J. Chem. Phys.* **30**, 648.
MOFFITT W. and YANG J. T. (1956) *Proc. Natl. Acad. Sci. U.S.* **42**, 596.
MOFFITT W., FITTS D. D. and KIRKWOOD J. G. (1957) *Proc. Natl. Acad. Sci. U.S.* **43**, 723.
MOSCOWITZ A. (1960) In C. DJERASSI (ed.), *Optical Rotatory Dispersion,* McGraw-Hill, New York, chap. 12.
MYERS D. V. and EDSALL J. T. (1965) *Proc. Natl. Acad. Sci. U.S.* **53**, 169.
PAULING L. and COREY R. B. (1951) *Proc. Natl. Acad. Sci. U.S.* **37**, 729.
PAULING L., COREY R. B. and BRANSON H. R. (1951) *Proc. Natl. Acad. Sci. U.S.* **37**, 205.
PELLER L. (1959)*J. Phys. Chem.* **63**, 1194, 1999.
PERUTZ M. F., ROSSMANN M. G., CULLIS A. F., MUIRHEAD H., WILL G. and NORTH A. G. T. (1961) *Nature* **185**, 416.
PYSH E. S. (1966) *Proc. Natl. Acad. Sci. U.S.* **56**, 825.
RAMACHANDRAN G. N., SASISEKHARAN V. and RAMAKRISHNAN C. (1963) *J. Mol. Biol.* **7**, 95.
ROSENFELD V. L. (1928) *Z. Physik* **52**, 161.
RUTTENBERG M. A., KING T. P. and CRAIG L. C. (1965)*J. Am. Chem. Soc.* **87**, 4196.
SAGE H. J. and FASMAN G. D. (1965) *Abstr. 150th Meeting ACS,* 107c.
SAMEJIMA T. and YANG J. T. (1964)*J. Mol. Biol.* **8**, 863.
SARKAR P. K. and DOTY P. (1966) *Proc. Natl. Acad. Sci. U.S.* **55**, 981.
SCHELLMAN J. A. and ORIEL P. (1962)*J. Chem. Phys.* **37**, 2114.
SCHELLMAN J. A. and SCHELLMAN C. (1964) In H. NEURATH (ed.), *The Proteins,* vol. 11, Academic Press, New York and London, chap. 7.
SCHERAGA H. A., SCOTT R. A., VANDERKOOI G., LEACH S. J., GIBSON K. D., OOI T. and NÉMETHY, G. (1967) In G. N. RAMACHANDRAN (ed.), *Conformation of Biopolymers,* Academic Press, New York, p. 43.
SHECHTER E. and BLOUT E. R. (1964a) *Proc. Natl. Acad. Sci. U.S.* **51**, 695.
SHECHTER E. and BLOUT E. R. (1964b) *Proc. Natl. Acad. Sci. U.S.* **51**, 794.
SHECHTER E., CARVER J. P. and BLOUT E. R. (1964) *Proc. Natl. Acad. Sci. U.S.* **51**, 1029.
SHMUELI U. and TRAUB W. (1965)*J. Mol. Biol.* **12**, 205.
SIMMONS N. S. and BLOUT E. R. (1960) *Biophys. J.* **1**, 55.
SOGAMI M., LEONARD W. J., Jr. and FOSTER J. F. (1963)*Arch. Biochem. Biophys.* **100**, 260.
TINOCO I., Jr. (1961) *Advances in Chemical Physics* **4**, 113.
TINOCO I., Jr. (1964)*J. Am. Chem. Soc.* **86**, 297.
TINOCO I., Jr., WOODY R. W. and BRADLEY D. F. (1963)*J. Chem. Phys.* **38**, 1317.

TODD A. (1960) In P. ALEXANDER and R. J. BLOCK (eds.). *A Laboratory Manual of Analytical Methods of Protein Chemistry*, vol. 2, Pergamon Press, Oxford and New York, chap. 8.

TOMIMATSU Y. and GAFFIELD W. (1965) *Biopolymers* **3**, 509.

TOMIMATSU Y., VITELLO L. and GAFFIELD W. (1966) *Biopolymers* **4**, 653.

TOWNEND R., KUMOSINSKI T. F., TIMASHEFF S. N., FASMAN G. D. and DAVIDSON B. (1966) *Biochem. Biophys. Res. Commun.* **23**, 163.

URNES P. (1964) Ph.D. Dissertation, Harvard University.

URNES P. (1965) *J. Gen. Physiol.* **49**, 75.

URNES P. and DOTY P. (1961) *Advances in Protein Chemistry* **16**, 401.

URRY D. W. and DOTY P. (1965) *J. Am. Chem. Soc.* **87**, 2756.

VELLUZ L. and LEGRAND M. (1965) *Angew. Chem.* **77**, 842; *Angew. Chem. Intern. Ed. Engl.* **4**, 838.

VELLUZ L., LEGRAND M. and GROSJEAN M. (1965) *Optical Circular Dichroism*, Verlag Chemie and Academic Press, Weinheim/Bergstr. and New York and London.

WADA A., TSUBORI M. and KONISHI E. (1961) *J. Phys. Chem.* **65**, 1119.

WILSON E. B., Jr. (1952) *An Introduction to Scientific Research*, McGraw-Hill, New York.

WOODY R. W. (1962) Ph. D. Dissertation, University of California, Berkeley.

YAMAOKA K. K. (1964) *Biopolymers* **2**, 219.

YANG J. T. (1965) *Proc. Natl. Acad. Sci. U.S.* **53**, 438.

YANG J. T. (1967a) In G. D. FASMAN (ed.) *Poly-α-Amino Acids*, Marcel Dekker, New York, chap. 6.

YANG J. T. (1967b) In G. N. RAMACHANDRAN (ed.) *International Symposium on Conformation of Biopolymers*, Academic Press, New York, p. 157.

YANG J. T. and DOTY P. (1956) Unpublished data.

YANG J. T. and DOTY P. (1957) *J. Am. Chem. Soc.* **79**, 761.

YANG J. T. and McCABE W. J. (1965) *Biopolymers* **3**, 209.

YANG J. T. and SAMEJIMA T. (1963) *J. Biol. Chem.* **238**, 3262.

ZIMM B. H. and BRAGG J. K. (1959) *J. Chem. Phys.* **31**, 526.

ZIMM B. H., DOTY P. and ISO K. (1959) *Proc. Natl. Acad. Sci. U.S.* **45**, 1601.

3

ULTRASENSITIVE REACTION CALORIMETRY

By T. H. Benzinger

from

Bio-Energetics Laboratories,
Naval Medical Research Institute,
Bethesda, Maryland

CONTENTS

3

ULTRASENSITIVE REACTION CALORIMETRY

By T. H. BENZINGER

from

Bio-Energetics Laboratories,
Naval Medical Research Institute,
Bethesda, Maryland

INTRODUCTION

Calorimetry has been used by chemists to investigate reactions over a long period. Yet ultrasensitive reaction-calorimetry applicable to proteins and to other macromolecular systems is so recent that few examples can be found in the literature. At this time the potential of ultrasensitive reaction-calorimetry in protein research must be derived from first principles. The liberation of heat as a non-specific, quantitative indicator of chemical change has no parallel in other phenomena that have been exploited for chemical analysis. These are more or less specific for certain reactions. Calorimetry is not. It has conceptually the potential of covering those areas for which no other analytical methods exist. While the non-specific nature of heat demands certain precautions, specificity can be provided by experimental design. Examples of such analytical applications will be given in this chapter.

Our second topic will be the application of ultrasensitive calorimetry to thermodynamic problems, directly to the heat of reaction as one of the two driving forces of chemical change in classical definition, and indirectly to the other driving force, entropy of reaction at finite temperature. Calorimetric determinations of chemical equilibrium make such an analysis possible. The principle of the method, its instrumentation and experimental procedures will be described first, then its analytical as well as thermodynamic applications.

A. INSTRUMENTATION

1. The Heatpulse Principle of Calorimetry

The classical objectives of calorimetry (beginning with the thermodynamics of fuels, explosives or fertilizer-synthesis) had posed no problems concerning amounts of substances and quantities of heat available for measurement. The aim was to determine enormous heats, with high precision. The accomplishments of classical calorimetry in the latter respect (accuracies within a few hundredths of 1%) were truly admirable. However, for classical measurements of heat of combustion, 10,000 gram calories were required. Two hundred gram calories were considered a suitable amount of heat for conventional reaction-calorimetry.* For protein analysis, measurements of a few thousandths of one gram calorie are often required.

The classical techniques are based on ultrasensitive measurements of temperature in large, known heat capacities under adiabatic conditions, which prevent losses of heat to the outside. With rare and therefore remarkable exceptions, for example Swietoslowsky[1] and Sturtevant,[2] classical principles of calorimetry were not applied to microanalytical situations.

When attempts are made to reduce the quantities of heat and material for measurement, by several orders of magnitude, the classical approach to the measurement of heat by ultrasensitive thermometry is faced with unusual difficulties. The surface-to-volume ratio of the sample rises. The external thermal disturbances which affect the surface become ever larger in relation to the small heat that is to be observed. Moreover, the thermal conductance of the devices for temperature measurement makes itself increasingly felt as a source of error. The general rule applies that with decreasing size of the object the very process of the observation begins to distort the object or event to be observed. An undesirable and unique characteristic of heat is experienced, namely that it leaks through any boundary, even through the barrier of empty space which it overcomes by radiation. Quite unlike the material objects which the chemist is accustomed to manipulate, heat cannot be confined in a test-tube. The requirement to confine the heat is circumvented in calorimetric methods based on heatflow.

Characteristically, two highly successful approaches to calorimetry of small amounts of heat were based on measurements of heatflow rates rather than temperature-changes. A. V. Hill[3] observed extraordinarily small and rapid changes of heat in nerve or muscle fibers with swiftly responding plated thermopiles using the principles of Wilson.[4] Hill's methods on the one hand and pulse-calorimetry, the topic of this article, on the other, are comparable in their emphasis on speed of response at a given heat capacity of the object. Also, both of these methods are capable of observing continuous heatflow rates of a few microcalories per second. However, Hill's objects were highly organized structures which are not reproducible, unlike the uniform populations of chemical species in

---

*Personal communication from the National Bureau of Standards.

solution. For the latter to be treated in a manner fitting their reproducibility new and stringent requirements for quantitative response must be met and new difficulties must be overcome. These arise from the disproportionately large heat capacity introduced by the solvent and from the need for mixing of reactants without undue thermal disturbance.

Another successful approach to microcalorimetry from heatflow rate measurements — not ultrasensitive thermometry — was made by E. Calvet.[5] His stated objective was "to measure accurately very small heat outputs over long periods of time". The voltage response of Calvet's instrument to heatflow is comparable with that of Hill's devices, and of the heat pulse calorimeter described in this article. All three methods are capable of measuring continuous heat production in the range of microcalories per second. Calvet's techniques were appropriately applied to slow inorganic absorption-processes, to the chemistry of cements, to the thermogenesis of highly complex biological systems and in other similar situations.

Pulse-microcalorimetry[6,7] differs from Calvet's method by its emphasis on speed of response which is exploited for sensitivity in measuring small amounts of total heat. Whereas the classical adiabatic approach tends to *decelerate* to zero rate the loss of heat from the reactants, the heatburst principle calls for the highest attainable *acceleration* of heatflow from the reactants through a thermopile into a heat-sink. Acceleration of the heatflow is, therefore, the principal technical problem. Finding new ways for the initiation of reactions without undue thermal disturbance proved to be equally important and difficult.

Pulse-microcalorimetry has been developed for two different, main objectives:(i) to exploit the unique characteristics of heat as a quantitative indicator of chemical change for inorganic, organic and biochemical microanalysis in science and technology; and (ii) to measure, in reactions of chemical or biochemical significance, heat and entropy changes, thermodynamic data of the chemical transformations. A further objective was to carry out these tasks in the millicalorie, or sub-micro-mole-range as routine procedures, with a standard instrument that can be made available to every laboratory for research or teaching. The objectives as outlined require the measurement of small total quantities of heat, where there is no continuous rate of production. To solve this problem, one further basic step had to be added to the principles of microcalorimetric method.

The total heat released from a chemical reaction under adiabatic conditions is represented by an elevation of temperature in the reactant system. In order to be measured as a flux of energy this heat must be discharged through a measuring device into a sink of practically infinite capacity which remains at the same temperature as the system before the reaction occurred. The heatflow can be made slow and long-lasting, or short-lasting but intense, depending on the conductance of the pathway. While there are practical limitations, there is no conceptual limit to increasing the rate of flow while the duration is proportionately shortened.

To exploit the energy available for measurement the heat pulse-principle of calorimetry calls for discharge of the given total heat into a sink, at the fastest attainable flowrate through a device that converts the flux of thermal energy into measurable electric energy. In this way useless "time of duration" can be exchanged for valuable intensity of the electric signal. For the maximum voltage obtained is proportional to the rate, not to the total of heatflow. After a sudden liberation of heat in the system, the rate of heatflow and the concomitant electric potential rise instantaneously and then decline in near-exponential fashion. The measurement of time presents no problem in this range. Shortening the duration of the flow is advantageous. It minimizes disturbances from environmental changes of temperature and flows of heat, as it shortens the duration of their effects.

The task then is to accelerate the flow of heat by widening and shortening of the pathway along which it is conducted. To accomplish this the reactant liquids are spread out as a thin layer over the largest attainable surface area. The large surface area of the liquids is the cross-section of the pathway of heat-conduction. The thickness of the layer of liquids is the length of the pathway through which heat is conducted to the thermopile. Through the thermopile the heat continues to flow, without a change in the large cross-section of its pathway, into the heat sink, again over an extremely short distance.

It is most advantageous to let the layer of reactant liquids form a "lining" of a cavity, inside the sink, rather than a layer on its outer surface (Fig. 1A). When a "four pi" closed cavity is lined with a thermoelectric

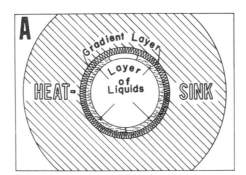

FIG. 1. Pulse-microcalorimeter, principle sketches
A. The relative locations of reactant liquids, thermocouple-gradient-layer and heat-sink.

"gradient layer" as in "Direct Calorimetry by Means of the Gradient Principle"[8] or in "Four Pi-Radiometry",[9] no heat can escape from it without being recorded. Figure 1A shows in cross-section a general outlay for a pulse-microcalorimeter. The cylindrical shape for the "lining" which the reactants form on the inner wall of the cavity in the block is advantageous for several reasons. As the layer of liquids returns into itself in a circle, the liquids move more easily and smoothly for mixing of the reactants, in vessels of annular cross-section. Cylindrical shape also facilitates the construction of the thermoelectric device which is the most important part in the task of building the instrument.

2. Area Thermopiles

Between the inner wall of the block and the outer wall of the reaction vessel (Figs. 1 A, B, C), a narrow space is left for a cylindrically curved, two-dimensional layer, a carpet-like structure, the thermoelectric device

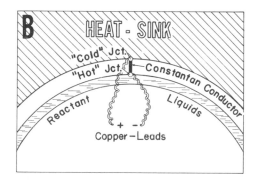

FIG. 1. Pulse-microcalorimeter, principle sketches
B. Single thermoelectric junction producing a potential when heat is liberated in reactant liquids.

that measures the heatflow rate which is proportional at any moment to the difference in temperature between the outer surface of the vessel and the inner surface of the sink. The thermopile, or "gradient layer", which measures this difference in temperature, consists of a multitude of alternating short, paired, serially wired, "conductors" for both heat and electricity, made from two thermoelectrically different metals, say, copper and the copper–nickel alloy, constantan. These "conductors" are extending radially between the outer surface of the reaction vessel and the inner surface of the sink.

It is apparent that the multiple conductors are arranged in parallel for the flow of heat from the reactant system into the sink, and in series for electric potential which they generate and send into an amplifying and recording system in response to heatflow. Looking at any single one of these approximately 10,000 paired conductors of heat and electricity (Fig. 1B) one readily sees that an electric potential arises when there is a difference in temperature between the "hot" junction adjacent to the reaction-vessel and the "cold" junction adjacent to the heat-sink. The same difference in temperature is translated into a difference of electric potential, instead of only once, in thousands of places (Fig. 1c) with thermoelectric conductors for heat and electricity filling the entire, narrow annular space between the vessel and the sink. Since these individual units are wired in series like galvanic cells in a battery, the numerous small potentials add up to a considerable voltage. The flow of heat through the numerous short, parallel conductors is swift and short-

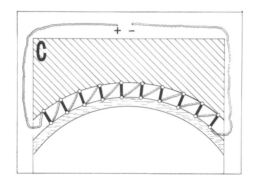

FIG. 1. Pulse-microcalorimeter, principle sketches
C. Arrangement of copper and constantan conductors in parallel for heatflow and in series for electric potentials.

lasting. The electric potential is high, because of the additive, serial arrangement.

As in the methods of gradient calorimetry and four-pi-radiometry, random-uniformity of the repeated structure without systematic errors must ascertain that when the thermoelectric response to heatflow is linear (as it surely is over the range under consideration, 0.0003°C in a measurement with 5 mcal total) it is also uniform, and independent of any irregularities in the local distribution of heatflow. The total coverage of the reactant system with thermoelectric conductors in a cavity ascertains, furthermore, that practically all of the heat generated is utilized for the measurement. Whatever losses take place, they will occur in the same quantitative relation to the total, during calibrations as well as during measurements. Also identical in calibrations and in measurements are the proportional, very small fractions of the measured heat which fail to be discharged into the sink because its capacity, though very large, is not infinite. To avoid errors from this source, heat capacities of liquids in identical, paired vessels must be kept nearly equal in measurements and calibrations. When this condition is fulfilled the measurements become independent of the heat capacity of the paired vessels and their contents.

3. Coiled-helix Thermopiles

The actual construction of the thermopiles for an ultrasensitive heat-pulse calorimeter presents considerable problems. To solder or weld in a uniform manner tens of thousands of junctions as required would be a difficult task. A satisfactory solution of this problem is the "coiled helix thermopile" (Fig. 1D) constructed as follows: upon flexible, in-

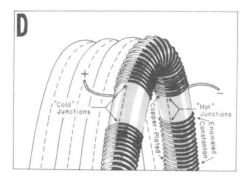

FIG. 1. Pulse-microcalorimeter, principle sketches
D. Apex of a coiled helix thermopile made by coiling of a coil of constantan-wire, half-copper-plated. Hot-junctions envelope the reaction-vessels, cold-junctions touch the hollow heat-sink.

sulating tubing of 3 mm o.d., temporarily filled with a fitting brass wire, an helix of enameled constantan wire of 0.15 mm o.d. is tightly wound. The outer part of the enamel-coating is then removed from one-half of every turn of the helix in such a way that two straight rows of boundaries become "hot junctions" and "cold junctions" when copperplating of the bared half (Fig. 2A) produces in the helix a continuously repeated alternating sequence of short, parallel, semicircular conductors of constantan and copper, plated upon constantan. As Wilson[4] found in 1917, a repeated structure with plated parts acts like a sequence of soldered links of constantan and copper in a conventional thermoelectric pile. The relative electrical conductance of the constantan wire underlying a substantial plated layer of copper is not sufficient to constitute a major short-circuit. The losses in potential are small compared with the advantages of this design.

The special microcalorimeter helix shown in Fig. 2A is coiled upon flexible non-conductive material such as polyethylene or teflon tubing. It can therefore be wound to form a cylindrical sheet or blanket-like structure, a gradient-layer surrounding the cylindrical reaction-vessel (Fig. 2B). For this purpose the wire-core is replaced with a thin steel wire as a mandrel, and the helix is coiled, in closely adjacent (though electrically insulated) turns, upon a thin-walled, cylindrical, electrically isolated bushing of metal into which the double-walled cylindrical reaction-vessel fits like a piston. Grease is applied to reduce the thermal resistance of the minimal air gap between the vessel and its bushing, but contrary to expectation the effect of this expedient on the speed of response is only minor. Fortunately, reaction vessels can be made not only of metal but also of glass with narrow tolerances (± 0.04 mm) using hand-drawn tubing. Two masters of the art of glass-blowing, F. A. Wild (Pasadena, California) and A. Leemans (Oxford, England), have perfected the reaction cells described in this paper.

With approximately 5000 junctions in close thermal contact with the metal bushing, electrical insulation of the helix becomes, of course, a serious problem. It is solved by anodic formation of a thin film of aluminumoxide on the surface of the bushing before the helix is wound. Likewise, a film of oxide is formed on the inner surface of the cavity in the aluminum sink before the helix is enclosed in it. While the helix is wound on the bushing it must be prevented from twisting, in order to achieve the optimal position of all thermocouples. The inner row of "hot" junctions is always kept oriented toward the outer surface of the bushing, as the helix is wound under considerable traction exerted by means of the mandrel of thin steel wire. The "cold" junctions are now facing outward, ready for thermal contact with the electrically insulated inner surface of the hollow heat-sink. The sink, a heavy cylindrical block of aluminum with a central bore, is split into three 120° sectors and forced upon the coiled-helix thermopile with close tolerance (Fig. 2C). The resulting compression of the resilient coils provides a uniform thermal contact between the "cold" junctions of the helix and the sink, as well as safe electrical insulation.

4. Twin-arrangement for Pulse-microcalorimetry

One coiled-helix thermopile around one reaction-vessel (Fig. 2B) when enclosed in a copper block would be an ultra-sensitive and rapid device to translate a sudden liberation of reaction-heat in the vessel into an electric potential change. However, the two criteria of voltage response and speed do not by themselves fully determine the performance of a microcalorimeter. A third, and equally important, factor is the indifference of the instrument to external and internal heat changes which are unrelated to the heat of reaction. Inside a pulse-microcalorimeter, heatflows arise when the liquid layers of stratified temperature undergo turbulent movements for the purpose of mixing, combined with the production of frictional heat and transitory changes of pressure. From the outside, the thermal balance of the instrument is endangered to a considerable extent even after extremely careful insulation. Therefore, in order to fully exploit the advantages of the heat-pulse principle the time-honored principle of twin-calorimetry (Pfaundler, 1891) must be utilized in this new connection. Into the heat-sink is placed not only one calorimetric system but two thermopiles and two reaction cells, both of identical construction in symmetrical locations, tandem-fashion (Fig. 2D). In one of the cells the reaction takes place. The other is a blank, containing the same quantities of liquids. Any generation or absorption of heat due to temperature-changes of the sink, or due to movements of gases and liquids in the cells when the instrument is tumbled or rotated, occurs with nearly identical intensity in both systems and creates thermoelectric potentials of virtually the same magnitude. These potentials cancel when the two identically constructed thermopiles are wired in series, with opposing polarity, in the same electrical measuring circuit. The heat of reaction is thus the only heat that is measured, with minimized disturbances from other sources. The quantity of the remaining disturbances is recorded as a "zero spike" of potential and will be described later.

5. The Heat-sink and its Insulation

Although the twin-arrangement largely compensates for changes of sink temperature it is still imperative to reduce, as far as possible, changes in average sink temperature as well as gradients of temperature throughout the sink. Two major approaches are available: (a) increasing the size and capacity of the sink and (b) insulating the sink against environmental temperature changes. Figure 2F illustrates how the capacity of the sink is reinforced by addition of a pair of heavy outer aluminum caps to the central part of the sink. These caps serve also to provide a thermal short-circuit between the three 120° sectors of the longitudinally split internal part of the sink (Figs. 2 D and E). Likewise, the three inner sectors extending over both the reaction and blank-thermopiles (Fig. 2C) provide the needed thermal short-circuit between the two caps as they form a bridge across their equatorial separation.

Size and weight of the block as shown must be kept within limits to permit the necessary swift accelerations and movements for mixing and to avoid the need for supporting structures of excessive weight. Insulation, the second approach, is therefore most important. Suspension of the sink in vacuum, theoretically the ideal solution, would introduce new difficulties in sealing the reaction-vessels, while it would yet require some means of thermally conductive, mechanical suspension for the block in the vacuum. It is, therefore, more advantageous to keep the sink suspended in air under atmospheric pressure and to protect it by means of two symmetric shields of Dewar-vessels (Fig. 2G).

These surround and enclose the sink almost completely from both sides. They leave between their vacua only a narrow circular gap of air and thermally conductive structures, namely, the glass necks of the two Dewar-vessels and three steel wires on which the heavy sink is suspended (Figs. 2 C and E). The three radial suspension wires originate at 120° angles, pointing outward to a mechanically strong equatorial ring. Strung with considerable traction these wires carry the sink by means of their tensile — not shear — strength, with minimal thermal conductance. Disturbing heatflows, either through the necks of the Dewar-vessels, or through the suspension-wires, arrive at the block in the equatorial plane of symmetry. They affect the twin thermopiles with equal intensity. The potentials thereby introduced cancel in the electric curcuit. The ring serves, furthermore, as a carrier for added outer twin-shields of alternating layers of insulating material and metal shells of substantial capacity as frequently used for thermal protection in calorimetry (Fig. 2H). By means of the ring — which may be tumbled in two bearings of a rotating yoke (Fig. 2H) — the entire instrument is rotated around its cylindrical axis, or tumbled end-over-end around the transversal axis, which is perpendicular to the axis of the cylinder and to the direction of gravity. These movements of the entire system are essential as the only permissible means of mixing reactants with an instrument of such sensitivity, without the aid of any moving or penetrating parts. Mixing of reactants in this manner depends, of course, upon appropriate design of the reaction-vessels.

6. Reaction-vessels

In a pulse-calorimeter the chemical reactants are accommodated in removable cylindric cells. These fit, when properly lubricated, like pistons into the bushings upon which the coiled-helix thermopiles are wound, with close tolerance to ensure the integrity of thermal conductance in the system. The interchangeability of the vessels permits the use of one standard instrument for any desired variety of thermal measurements. With various choices of cells of the same standard outer dimensions, but different internal devices for compartmentalization (Figs. 3 A–J, 4), the instrument may be adapted to interactions of solids, liquids or gases, to experiments with or without vapor space, and to measurements with any

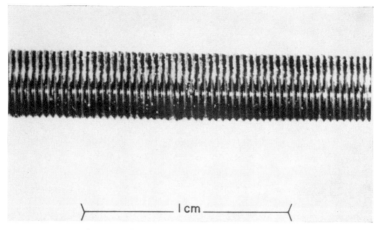

FIG. 2. Pulse microcalorimeter: actual construction. A. Flexible, plated thermocouple-helix, enlarged. Note the bright overlay of crystalline, plated copper on the upper half of the coil. Row of thermoelectric junctions runs along centre line.

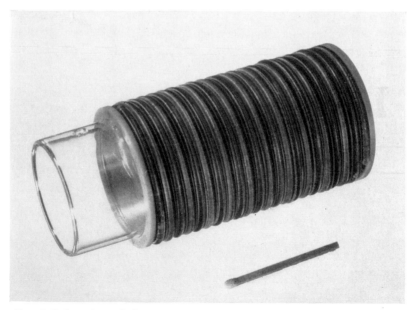

FIG. 2. Pulse microcalorimeter: actual construction. B. Flexible thermocouple-helix coiled upon metal bushing with insulating flange, with annular reaction-vessel introduced part of the way into the bushing. Note the insulating spacers between the turns of the helix.

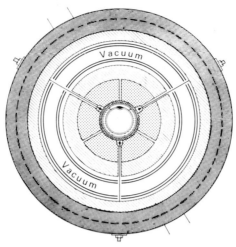

FIG. 2. Pulse microcalorimeter: actual construction. C. Equatorial (off-center) section through the instrument showing a reaction-cell, a thermopile, the 120° trisected sink with a circular cap and an evacuated Dewar-vessel. The suspension-wires (shown as though the section were centered) extend outward to the suspension-ring.

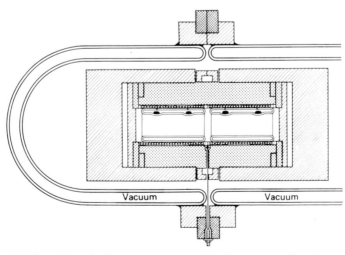

FIG. 2. Pulse microcalorimeter: actual construction. D. Longitudinal section showing disposition of paired vacuum vessels, composition of the heat sink and axles for tumbling (outer shells and axles for rotation are not added in this drawing).

FIG. 2. Pulse microcalorimeter: actual construction. E. Block suspended in ring.

FIG. 2. Pulse microcalorimeter: actual construction. F. Caps added to reinforce
the heat-sink.

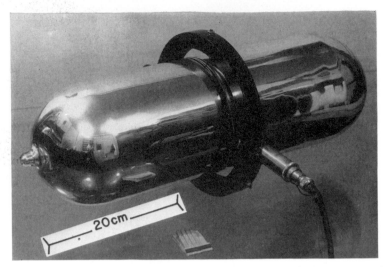

FIG. 2. Pulse microcalorimeter: actual construction. G. Paired Dewar-vessels surrounding the core of the instrument.

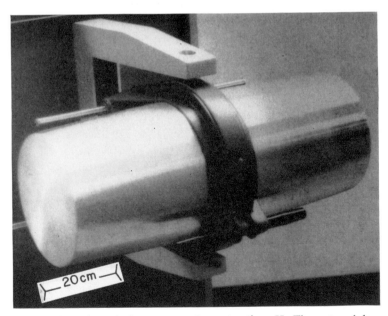

FIG. 2. Pulse microcalorimeter: actual construction. H. The external heavy insulation is fastened to the suspension ring. Handles serve to let outer shells slide and move off-center. Yoke permits tumbling movements. Bearings in yoke receive axles of the suspension-ring for rotating movements. (Outer dimensions of ring are larger in Fig. H, a photo of the commercially available instrument. Figures E, F and G were taken with the laboratory-prototype of references (6) and (7).)

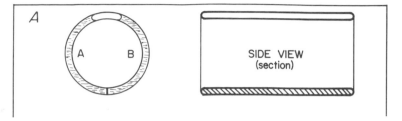

FIG. 3. Reaction-vessels for pulse-microcalorimetry, principle sketches
A. Bicompartmented vessel with longitudinal partition at bottom.

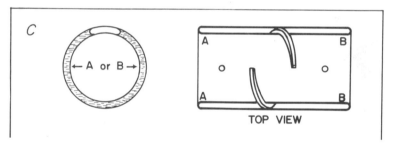

FIG. 3. Reaction-vessels for pulse-microcalorimetry, principle sketches
B. Bicompartmented vessel with equatorial partition.

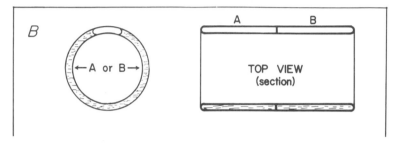

FIG. 3. Reaction-vessels for pulse-microcalorimetry, principle sketches
C. Bicompartmented vessel with spiral partition.

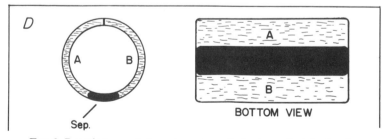

FIG. 3. Reaction-vessels for pulse-microcalorimetry, principle sketches
D. Vessel for experiments without vapor-space. Inert liquid forms longitudinal partition.

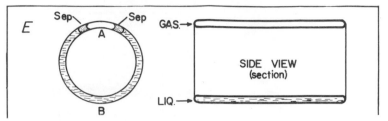

FIG. 3. Reaction-vessels for pulse-microcalorimetry, principle sketches
E. Vessel with inert liquid for separation of a liquid from a gaseous reactant.

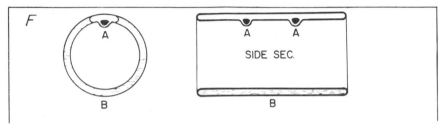

FIG. 3. Reaction-vessels for pulse-microcalorimetry, principle sketches
F. Standard "drop" vessel with dimpled recesses for one reactant in small amounts, other reactant in bulk space.

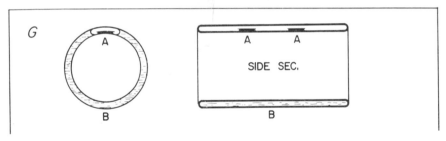

FIG. 3. Reaction-vessels for pulse-microcalorimetry, principle sketches
G. Drop-holders in metal vessels hold one reactant in narrow space between cylindrical surfaces.

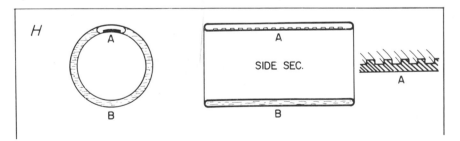

FIG. 3. Reaction-vessels for pulse-microcalorimetry, principle sketches
H. Open cylindrical grooves on surface of inner shell hold one reactant by capillary action.

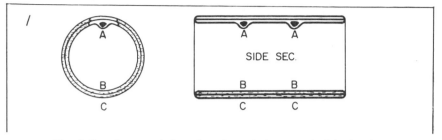

FIG. 3. Reaction-vessels for pulse-microcalorimetry, principle sketches
I.   Photochemical reaction-vessel admits the components of a light-producing
reaction in the inner, annular space. Photons produce a light-absorbing
reaction in the outer space of the vessel.

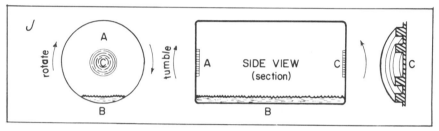

FIG. 3. Reaction-vessels for pulse-microcalorimetry, principle sketches
J.   Rotating vessel needed for biological material with gas metabolism and.ten-
dency for sedimentation. Two different agents can be added in sequence by
tumbling motions to initiate or inhibit reactions.

ratio of volumes of two, and occasionally more than two, reactants.
With one exception (Fig. 3J) the shape of these vessels is annular. The
vessels and the liquids which they contain offer only a circular edge, not
an appreciable surface to the two end-walls of the cylindrical cavity,
which are not covered with a gradient-layer. Aluminum foil discs, fastened to the open ends of the vessels, prevent most of the possible radiant
or convective losses of heat from the inner surfaces of the annular vessels.
However, when this precautionary measure is omitted in calibrations as
well as in experiments, no perceptible errors are observed.

   In vessels for similar volumes of two reactant liquids, the two participants are kept apart by their gravity in spaces divided by a partition
(Figs. 3 A and B). When the vessels are tumbled, the liquids are exchanged
and mixed through wide openings between the compartments, located in
those parts that are air-filled before and after the tumbling movements.
In drop-type vessels one reactant liquid is present in small quantity (for
example, 100 to 10 μl or less) whereas the other reactant fills the bulk-
space (5 to 15 ml). Drops may be held and kept separated from the bulk,
in dimpled wells or recesses, by gravity (Fig. 3F). In other types of cells

capillary attraction serves to hold reactants separated in narrow spaces between plane or cylindrical surfaces (Fig. 3G), in open grooves on the inner surface (Fig. 3H) or between convolutions of filaments of suitable material in the air space. All drop-type vessels are mixed by simple rotation of the instrument without tumbling.

In most of the vessel-types, two countersunk filling-holes of 2 mm diameter serve to introduce the reactants from pipettes, micro-pipettes or syringes through polyethylene tubings. The filling-holes are located at the apex of the outer shell. They are sealed with lubricated plugs of resilient plastic. The plugs are shaped like countersunk screws and they must not protrude over the cylindrical, fitting surface of the vessel.

To facilitate the swift conduction of heat and to avoid encumbrance with additional heat capacity, the vessels are made as thin-walled as possible. Cells of glass or teflon are best suited for biochemical reactions in which heavy metal ions might act as enzyme poisons. Metal vessels, on the other hand, have the advantages of high thermal conductivity and structural strength. They can be taken apart for cleaning purposes and are, nevertheless, well sealed when assembled. For experiments that involve drastic dilutions of electrolytes, noble metal vessels are indispensable. On non-conductive solid–liquid interfaces with ionic solutions, electric charges are not reproducibly distributed. A change in their distribution, which occurs when concentrations of the ionic solutions change, causes a liberation or absorption of heat. With vessels of noble metal, this phenomenon does not arise.

In other experiments, where no vapor space is desired, inert liquids heavier than the reactant liquids (for example, mercury) may be used to keep reactants separated (Fig. 3D). A top layer of inert liquid lighter than the reactant may be considered for temporary separation of a gas from a reactant liquid before mixing (Fig. 3E). Solids may be kept in drop-wells (Fig. 3F) before interaction with liquids, for example, in measurements of heat of solution or absorption. A special problem arises when photons are a reaction-partner, as, for example, in experiments on photosynthesis. Optical devices for introduction of light into the calorimeter would invite serious problems of generation, conduction or storage of heat. A different design (Fig. 3I), developed in collaboration with L. Kiesow, permits the use of photons from a light-generating chemical process (such as luciferin–luciferase–ATP–interaction) in the inner space of a triple-walled vessel as a participant of a light-absorbing reaction in the outer space. Figure 4 shows the actual vessels, embodiments of the schematic drawings of Fig. 3.

Recent additions to existing types of vessels permit a more rapid response and, therefore, increased power of resolution with liquid layers of one instead of 3 mm thickness and a total capacity of 5 instead of

15 ml. Figures 5 and 7A show a drop-type and a bicompartmented vessel of this kind.

One rather serious limitation of the method—now overcome—was presented by the requirement for diffusion of respiratory gases into, and out of, suspended biological material. In the techniques of Warburg[10] this well-known complication of studies on cell metabolism has been resolved by continuous agitation of the manometric reaction-vessels. In microcalorimetry the difficulties are compounded by the need for a continuous, undisturbed exchange of heat between reactant liquids, vessel, thermopile and block. An efficient device is the vessel shown in Fig. 3K, which permits continuously repeated rotating movements for exchange of gases (+ 180° forward, 180° return) without mixing of liquids. Five milliliters of solution in this vessel produce an elongated body of liquid in contact with whatever sector of the cylindrical thermopile is facing downward. During rotation, the bulk of the liquid is aerated and moved with respect to the thermopile and sink. In this way a transient and repeated thermal contact of the liquid with all sectors of the thermopile is established. The small and regular oscillations of potential caused by these movements (Fig. 6A) do not disturb the measurements. The heat of friction, whatever its magnitude, is compensated satisfactorily by the twin-arrangement. A second and third reactant are kept separated from the bulk in narrow, circular grooves, at the central regions of the discs which close the two ends of the cylindrical vessel. When thermal equilibrium is attained, these drops are added to the bulk by tumbling, first to the right then to the left, in separate actions. Thereby, the grooves are swept clean of the reactant which they contained. Figure 6A, recorded with a stainless-steel vessel of the type Fig. 3J, shows the first measurement of two consecutive reactions in one vessel, observed without opening of the calorimeter. This type of vessel extends the potential of pulse-microcalorimetry to interactions of chemical compounds with complex biological systems such as small metazoa, protozoa, tissue slices, dividing germ-cells, micro-organisms, cell-cultures, virus host-combinations, isolated cells, contractile structures, membranes, subcellular particles, or any other material which tends to sediment or has a requirement for the exchange of respiratory gases.

In summary, any new plans for applications thus far have left the heat-measuring device proper unchanged. The heat-pulse principle and the instrument lend themselves equally well to many different types of calorimetric research topics, simply by differences in design of the reaction-vessels. This situation differs markedly from the past efforts, where calorimetry often involved as many programs of instrument design as there were topics of investigation.

One special type of vessel still remains to be discussed. For electrical calibrations of the instrument, a precision resistor must be accommodated in a vessel of otherwise standard construction. Manganin wire, 36 gauge, is embedded in a sheath of thin polyethylene tubing, provided with terminal leads of copper, and coiled within the liquid-space of a vessel which is then filled with water as usual, but permanently sealed. The resistance of the calibration wire must be precisely known from comparison with a resistor certified by the National Bureau of Standards.

Calibration by means of water-fillted, not empty, vessels is important. The large heat capacity prevents substantial temperature increases in the heating wire which would result in losses of heat through the electric terminals.

7.   Accessory Equipment

For calibration and for use a pulse-microcalorimeter must be provided with conventional accessories: a direct-current amplifier with a maximum gain factor of 400-1000, a recording potentiometer, 10 mV full scale, and a ball-integrator combined with the recorder. (The automatic integration of the areas under the recorded curves is superior to manual integration because there is a human factor in the latter, as the planimeter-lever must be moved by hand along the recorded lines.) It is convenient to have available a number of lower steps of amplification. These permit measurements of heat in ranges above the millicalorie range by several orders of magnitude. Lower amplification permits, furthermore, observations of how the recorded potential approaches the baseline, when the instrument has been newly charged with reactants in two cells of nearly, but not fully, identical temperatures. A smooth, near-exponential approach is proof that no traces of water have been left on its cells, or has leaked from filling-holes. (The evaporation even of invisible traces of moisture produces large and erratic thermal potentials.) To extend the range of observations from room temperature or the thermodynamic standard temperature, 298°K, downward as far as 278°K and upward to 350°K, the instrument can be operated in a small, temperature-controlled enclosure. At any given temperature, the controlled enclosure serves to minimize environmental thermal disturbances.

8.   Calibrations

A heat-pulse microcalorimeter may be calibrated electrically, either by pulses of current or by steady inputs of current. Figure 6B shows an electric calibration with a steady current sent through a resistance-wire in a water-filled vessel. The instrument responds to a steady flow of current or heat with a steady potential, which rises and declines in exponential fashion upon initiation or termination of the current. In a calibration of this kind the voltage response of the calorimeter is conveniently expressed in $\mu V/cal/sec$.

Another way to calibrate the instrument is by introducing a pulse of known total heat. This can be done electrically (Fig. 6c) as well as by means of a chemical reaction, for example, with a pulse of heat from the instantaneous reaction

$$H^+ + OH^- \rightarrow H_2O$$

performed by mixing acid (0.8 μmoles HCl) and base (in excess) in the reaction vessel (1/10,000 M NaCl is added to avoid surface-interactions to be discussed later). The instrument responds to a pulse of heat with an almost instantaneous rise of potential followed by a near-exponential decline.

In chemical calibrations it is necessary to exclude the interaction of choride ions with the glass wall of the vessel by addition of sodium chloride in 1/10,000 molar concentration to all reactant solutions. Furthermore, a low molarity of reactants (0.01 M HCl in the drop-wells and 0.001 M NaOH in the main vessel space) is recommended to make heats of dilution negligible. H. Skinner* suggests instead the following procedure:

Cell A	Cell B
200 μl [0.01 M HCl/0.001 M NaCl]	400 μl [0.005 M NaCl]
15 ml [0.01 M NaOH/0.0001 M NaCl]	14.8 ml [0.01 M NaOH/ 0.0001 M NaCl]

so that the measured heat corresponds (without correction) to 200 μl [0.01 M HCl] + 200 μl [0.01 M NaOH] $\rightarrow$ 400 μl [0.005 M NaCl] + water.

However, this procedure still includes the heats of dilution of sodium and chloride ions from 0.01 to 0.005 molarity. We suggest to satisfy the conceptual requirements with one further control: heat of dilution of 0.01 M NaCl to 0.005 M with 0.001 M NaCl in a bicompartmented vessel. By subtraction of this heat from the heat observed in the neutralization experiment ΔH is obtained for the reaction.

$$H^+_{(aq., 0.01 M)} + OH^-_{(aq., 0.01 M)} = H_2O.$$

* Personal communication.

For the heat of this reaction at infinite dilution Harned and Owen (11) give values of

$$\Delta H_i = -13,721 \text{ cal/mole at } 18°C$$

and

$$\Delta H_i = -13,606 \text{ cal/mole at } 20°C.$$

The temperature coefficient of 57 cal/mole/deg must be considered in calibrations performed at different temperatures.

9. Performance Characteristics

The performance of the instrument depends with equal emphasis on two factors: (a) the voltage response to heatflow as described, and (b) the speed of response expressed as the half-response time of the exponential decay after an instantaneous heat-pulse. The speed of response determines the height of the spike of potential in response to a heat-pulse of given total and recorded area. Thereby, speed determines the practical usefulness of the voltage-sensitivity in measuring small total quantities of heat, not rates of heatflow. Speed also determines the power of resolution for reactions with mixed kinetics and for the time-course in heat changes in complex systems. In standard pulse-micro-calorimeters that can now be made available to every laboratory, the voltage-sensitivity of the thermopiles is of the order 200,00 μV per gram calorie in a second. The half-response time is of the order 100 sec. (With specially designed reaction vessels, half-response times as short as 30 sec have been attained.) These characteristics of performance make it possible to measure total heats of the order 5 mcal with an accuracy of $\pm 11\%$. The instrument is capable of detecting short-lasting and sudden changes of temperature by $3 \times 10^{-6}°C$ in the reactant liquid. Solutions of 1/20,000 molarity and fractions of 1 μmole total substance have been used in our investigations. Above the limiting range the calibration of the instrument holds unchanged through several orders of magnitude.

In obtaining these advantages for the measurement of small total heat with reactions of limited duration, advantages in measuring continuous heat output at very low rates need not be sacrificed. Long-term stability of the recorded baselines, a main requirement for the observation of low rates, has been achieved with a heat-sink of moderate size by protection with vacuum. The restriction in size of the sink, adopted to permit the mixing movements, allowed the use of glass Dewar vessels for protection. As a result, small instantaneous bursts of heat as well as the heat production of small continuous amounts and combinations of these two may be recorded. When chemical or living agents are added to metabolizing systems, instantaneous interactions as well as lasting changes of the rate of heat production may occur.

B. EXPERIMENTAL PROCEDURES

1. Electrical Preparations

Before experiments with the ultrasensitive calorimeter the amplifying and recording system must be carefully checked out. During this check, thermal influences are best eliminated by temporary replacement of the terminal leads from the calorimeter to the amplifier, with a resistor of the approximated resistance of the two calorimeter thermopiles. (The resistor and its connections must be well protected from differences or changes of temperature, to avoid disturbing thermo-potentials.)

Next, the amplifier inside the recorder that magnifies the error signal to which the servomotor responds must be adjusted to the contemplated range of voltage. The gain factor must be neither too high nor too low. Excessive sensitivity at high preamplification produces unnecessary noise. Insufficient sensitivity fails to activate the servomotor during slow changes of potential in the recording. It leads to step-changes where continuous slow changes of potential take place. Optimal sensitivity is recognized by a recorded line of approximately 1 mm band-width, in which the amplifier noise is clearly visible indicating that the system will respond to every change of potential without undue unertia.

The calibration of the amplifying and recording systems is confirmed by recording of a known standard voltage from a potentiometer with a certified standard cell together with appropriate lengths of zero-voltage baselines before and after the deflection. Equally important is certification of the transport-speed of the recording paper, which is checked with a stop-watch. The integrating device is checked for validity of its calibration against a recorded voltage difference and time-interval. For hand-planimetry a rectangle is drawn on recording paper with a known voltage-deflection as abscissae and a known time interval as the ordinate. Taking this area under the planimeter, one obtains a factor which permits for any recorded curve to translate planimeter reading units into "microvolt-seconds", the measuring unit for calories in the heatburst-instrument.

2. Perfection of Thermal Baselines

The precision of heatburst-measurements depends largely on the quality of the "thermal baseline" obtained when the thermopiles are connected with the amplifying and recording system while no electrical or chemical heat production or absorption takes place inside the instrument. This baseline (a) must not visibly deviate from straight-line-characteristics, (b) must be parallel to the electrical zero baseline, (c) should not be offset by more than 0.5 μcal/sec from electrical zero. With a baseline of such characteristics, recordings may be carried out in the range of 5 to 3 mcal, evolved in practically instantaneous reactions.

The thermal baseline cannot be judged from observations on the empty instrument. Water-filled reaction vessels (or metal dummies of similar heat capacity and dimensions) must rest in the thermopiles. The degree in which baselines are affected by changes of sink temperature at the "cold" junctions depends on the heat capacities adjacent to the "hot" junctions. Heat capacities for testing should therefore be similar to experimental capacities.

In order to obtain near-perfect baselines for calibrations as well as for experiments, the following rules should be observed. All operations should be carried out symmetrically. The insulating armor must be removed simultaneously from both sides; the vessels must be introduced, within a short time interval one after another; the plugs of the block must be inserted bilaterally; the insulating armor must be closed simultaneously from right and left. To avoid heat transfer the block must not be touched with bare hands. Vessels should not be touched when filled. During periods of adjustment, one should maintain that particular horizontal position of the calorimeter in which the reaction and blank vessel-poles of the instrument are facing the right and left walls, not the rear and front walls of the chamber. (Temperature differences between the rear wall facing the machinery for temperature conditioning, and the front wall facing the open room, might exert unilateral influence. Also, floor and ceiling of the room often have different temperatures.)

The quality of the thermal baseline depends not only on the prevailing thermal situation. It is a result of "previous thermal history". The block, when unilaterally heated, would transmit different temperatures to the insulation and to the glass walls of the Dewar-vessels. Because of the low thermal conductivity of those objects, the differences in temperature would persist. When chamber-temperature has been raised, a longer time is required for recovery of the baseline. During such waiting periods the instrument should be kept open, not closed.

When after closing of the instrument the baseline has returned to zero, the mixing movement should be performed only after a sufficient waiting period. Occasionally, a baseline going through a shallow minimum or maximum will slowly rise again, or decline before it settles.

When the amount of heat to be measured is very small, below 5 mcal, one should read the individual potential from the reaction-thermopile alone, to ascertain that a true thermal equilibrium has been obtained, not an apparent equilibrium, by compensation in the twin-thermopile arrangement. With an even distribution of temperatures throughout the block, zero spikes as a correction become smaller.

Last, not least, the temperatures of vessels and solutions before their introduction should differ as little as possible from the temperature of the block. They must not exceed block temperature because cooling in the block would lead to condensation of water vapor from the saturated air-space on the inner walls. Heat of dilution would be observed as an error, when reactants or ionic media are diluted by the water-condensate upon mixing.

3. Filling and Introduction of Reaction-vessels

In filling reaction-vessels the following viewpoints are important: (a) volumetric transfer of correct amounts, (b) keeping reactants separate before mixing, (c) prevention of the appearance of water on the surface of a vessel, (d) prevention of water penetration through filling holes, (e) prevention of the formation of liquid water-films on the inner walls of vessels, (f) prevention of pressure differences, (g) prevention of major deviations of vessel temperature from block-temperature. In the following sequence of procedures, these viewpoints are considered.

The first step before charging a vessel with reactants is application of the lubricant to its surface (Fig. 7A). If this were done after filling, absorption of heat from the lubricating fingers would be substantial because of the increased heat capacity. Until all spaces are filled, the filling holes and their surroundings remain exempted from lubrication. After lubrication the vessels are filled with the aid of various devices of flexible plastic, not glass. Chipping of filling-holes, or breakage of the thin-walled vessels, can thus be avoided. To fill the main spaces of drop vessels, or the compartments of bicompartmented vessel types, glass pipettes provided with polyethylene tubings, tapered at both ends (Fig. 7B) are applied. One taper fits into the outlet of the pipette, the other passes through the filling-hole into the depth of the main vessel space. Drop-wells are filled, either with constriction-pipettes of flexible plastic (Fig. 7C, top) or with polyethylene tubing of 0.024 in. o.d., attached to a hypodermic needle and microsyringe (Fig. 7C, right). Correct quantities can also be measured into vessels by a third way: filling of a coiled polyethylene tubing 0.024 in. o.d. with reactant, weighing it on a micro-balance, displacing the contents into the drop-well of the vessel, and reweighing the tubing. (The specific weight of the solution thus transferred must be known or be reasonably close to unity.) No traces of solutions must ever be left around filling-holes. A wick of fine tissue is used for removal if necessary. Now the filling-holes may be lubricated and the plugs, except for one at the outer end of the vessel, are put into place. Plugs must require a considerable force while being pushed into the hole, to obtain a satisfactory seal. If plugs fit loosely, they cannot be trusted.

Figure 7D shows how vessels are introduced into the calorimeter. They are left resting on the supporting block which is held by one hand, while the other hand gently pushes the vessel into the central bore, with a rod of insulating material. The vessel is first introduced partly, so that one filling-hole remains open. As soon as the potential from the instrument is on-scale of the recorder, the last stopper is pushed in for seal, with a blade or knife (held horizontally and flat to prevent the plug from being pushed sideways). Then the vessel is fully inserted, always to the correct position as recognized by the edge of the thermopile-sleeve where it borders the insulating end-ring.

Whenever the potential from the thermopiles fails to approach the baseline in smooth, near-exponential fashion, or is substantially off-set from zero, the reason is evaporation of minute amounts of water inside the instrument. (Compared with 5 mcal as the range of sensitive measurements the heat of evaporation of water is very large, more than 500 mcal/μl.) If the baseline is unsatisfactory the instrument must be reopened, the vessels removed, stoppers and filling-holes checked, the lubrication wiped off and renewed and the vessels reinserted after thorough cleaning of the interior of the block with gauze or tissue. It is seldom possible to find where the leak occurred. The amounts that cause the disturbance are too small. Prevention of water-leaks depends on the quality of the filling-holes and plugs and on a smooth cylinder-piston fit between the central bore of the calorimeter and the reaction vessels as a second, water-proof seal. Warming of vessels after sealing must be avoided as it leads to increased internal pressure, and may cause leakage when during mixing movements the vessels are inverted and water is adjacent to the filling holes.

4. Mixing Movements

Reactants are mixed after establishment of a desirable baseline, by rotation of the instrument around its longitudinal axis, or by tumbling, end-over-end. For drop-vessels a recommended standard sequence of rotation is: 180° clockwise, 360° counter-clockwise, 180° clockwise.

Bicompartmented vessels must be tumbled. In tumbling of vessels with equatorial partition a time interval (3 sec) must be allowed between individual turns so that liquids can flow from one compartment to the other through the narrow passage. A useful sequence for tumbling is: 90° clockwise — pause, 180° clockwise — pause, 270° counter-clockwise.

Thereafter the instrument is rotated 90° clockwise or counter-clockwise, in order to bring the liquids to identical levels in the two compartments. After 3 sec a 90° rotation in the opposite direction restores the initial position of the instrument. In bicompartmented vessels with longitudinal partitions equal volumes of compartments are maintained during tumbling movements without special precautions. However, identity of the initial and terminal positions is equally needed. To avoid baseline-shifts, the heat capacity of the liquids must "see" the same population of thermocouples before and after mixing. To avoid displacements of vessels in their thermocouple-sleeves — which would also cause baseline-shifts — accelerations and decelerations of the mixing movements must be very gradual, never abrupt.

Requirements for mixing-sequences depend strongly on other factors such as the viscosity of the liquid and the stoichiometric relations (excess) of reactants. For different individual situations optimum moving sequences must be found experimentally, and checked by identity of two zero spikes after termination.

5. Advantages of Using Dilute Reactant Solutions

In calorimetric experiments heats of dilution (unless they are the object of the study) should be minimized as a source of error. When other considerations permit, reactant concentrations should be as low as allowable with a given sensitivity of the calorimetric method. This emphasizes again the usefulness of highly sensitive instrumtation. It also emphasizes the importance of experimental design which eliminates by cancellation, heats of dilution of reactants and products (see Section D.1 on ΔH-determination).

When macromolecular systems are investigated, ionic media of low molecular weight should be dilute if possible, and carefully balanced by dialysis of the two reactant solutions in the same medium prior to calorimetry. Errors due to dilution of water-condensate at the walls of the gas-space of reaction vessels are likewise minimized by low concentrations of reactant solutions and ionic media.

Calorimetric determinations of chemical equilibria with reasonably slow reactions are not disturbed by dilution. A spike of instantaneous dilution-heat is readily distinguished from the slower heat of reaction. However, in this context, the problem of activity coefficients arises. Reactant solutions should be as dilute as possible, to minimize the influence of uncertainties concerning the activity coefficients.

6.   Advantages of Using One Reactant in Excess

The choice of relative amounts, not only concentrations of reactants, is important in ultrasensitive calorimetry. With one of two reactants in substantial excess reactions are complete, even if the mixing process was not perfect. Figure 8 demonstrates how a reaction in bicompartmented vessels may become incomplete in the range of stoichiometric equivalence. This range, if possible, should be avoided in experiments as well as in calibrations.

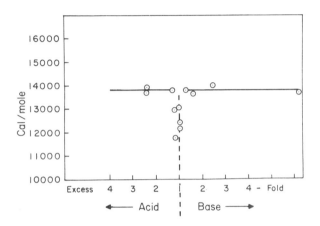

FIG. 8. Excess of one reactant. Excess of one reaction-partner ascertains completion of reaction in bicompartmented vessels. In this example the heat of formation of water was observed by mixing acid with base.

It would, therefore, be inadvisable to select as a standard for calorimeter-testing a heat of dilution; for example, the dilution of potassium chloride. With such a process the terminal state is attained only by perfect mixing. Moreover, drastic dilution of electrolytes in non-conductive reaction vessels leads to non-uniform distribution of electric charges and to disturbing heats of their redistribution. This phenomenon also argues against the potassium chloride standard.

Another class of measurements for which perfect mixing is an indispensable condition are determinations of chemical equilibria. Whenever possible these should be carried out in drop-type, not in bicompartmented vessels.

Special problems arise when one macromolecule accepts not one but many small molecules or ions. When these are added from dropwells at high concentration, those macromolecules that come in contact with them first accept a number larger than the final average.

Thereafter a slow process of redistribution takes place, delayed by the uneven distribution of macromolecules that have reacted with different numbers of small molecules or ions. Each round of agitation of the vessel accelerates the slow redistribution. Thus, total heat of reaction is liberated not at once, but in several, decreasing steps, until equilibrium is achieved. The sum-total of the individual heats observed is the heat of reaction. Similarly, determinations of heats of solution of solid compounds in the microcalorimeter may require repeated agitation and summation of the recorded heats. Identity of the two final "zero spikes" is the criterion for complete reaction.

7. Some Aspects of pH in Calorimetry

When two different compounds such as proteins in solutions of different pH are interacted in the calorimeter, heat of ionization is observed. From measurements of pH before and after the reactions in the known volume one finds the net number of protons bound or released. The heat of ionization can thus be measured, if no other interactions except ionization take place. If the object of the investigation is a protein–protein interaction, not ionization, then pH of known reactant solutions must be carefully balanced before the calorimetric experiment, as ionic contributions to heats of reactions may be substantial. In reactions where protons are a participant, the heat of ionization or neutralization of protons in the buffered medium is measured separately, by addition of protons (a fully dissociated acid) to the medium in a control experiment. Examples of this will be shown below.

8. Heat of Gas Metabolism

Living matter of more than macromolecular size tends to sediment in calorimeter vessels. When the metabolic heat production of such material depends on oxygen consumption, agitation of the vessel will accelerate it and produce a burst of heat. (Oxygen from the air space is then carried to the metabolizing particles by rapid convection instead of slow diffusion.) Therefore, continuous agitation in special vessels is a condition for this type of work.

Metabolic heat becomes a source of error when biological materials such as proteins or nucleic acids are infected with a microorganism. Since sterile techniques are not a standard requirement in protein work, this possibility must be considered, when heat evolution larger or longer-lasting than a zero spike is observed upon agitation of biological material.

9. Heat of Precipitation

Antigen–antibody interactions are usually recognized by the precipitation of an insoluble compound. Ultrasensitive calorimetry permits the detection of antigen–antibody interactions that do not form a precipitate. How wide the range of such unknown interactions may be is a matter of speculation. Whenever the reaction under study does form a precipitate, the heat of precipitation—often a multiple of the heat of interaction proper between antigen and antibody—must be established by additional measurements under conditions where precipitation does not occur, viz. in the range of antigen excess.

10. Dissociation

Compounds of the type AB that are dissociating into compounds A + B will perform the dissociation reaction whenever they are diluted, because in the dilute solution, equilibrium with

$$K = \frac{[A]\,[B]}{[AB]}$$

is restored by dissociation of AB. This effect must be kept in mind when dissociable compounds are used in calorimetric experiments. The heat observed upon interactions of such compounds with others may require a correction for the heat of dissociation, not only dilution as a result of mixing.

C. ANALYTICAL APPLICATIONS OF ULTRASENSITIVE REACTION CALORIMETRY: CONCRETE EXAMPLES

1. Calorimetric Demonstration of Unknown Interaction

An as yet unclarified reaction is chosen to demonstrate the application of calorimetry as a probing tool. It had been suspected that glutathione, by means of its SH-group, could form a new compound with glyceraldehyde. However, there was no method known for separation of such a compound, nor for measuring its quantity. The pulse-microcalorimeter demonstrated not only the reaction as such but also its reversibility. The simple expedient in finding the reversibility was repeating the reaction with three different concentrations of one reaction partner, D,L-glyceraldehyde. The heat, ΔH, of the unknown reaction does, of course, not depend on the concentrations of reactants or products. When less heat, 545 cal/mole of D,L-glyceraldehyde charge, was found

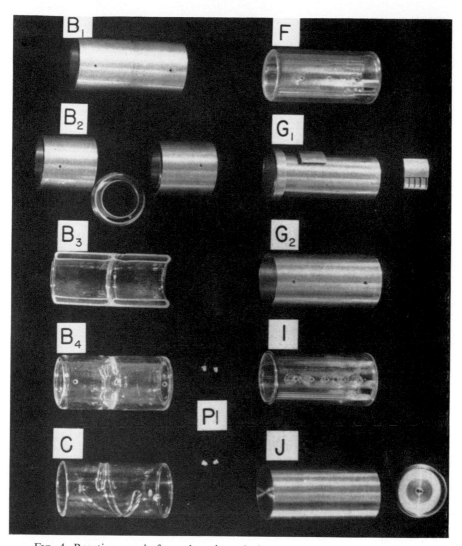

Fig. 4. Reaction-vessels for pulse-microcalorimetry. To the left a bicompart-mented cell of platinum–rhodium alloy is shown closed (B_1) and disassembled (B_2). Note equatorial ring and partition with kidney-shaped hole for mixing of reactants by tumbling movements. A bicompartmented glass cell is shown in section (B_3) and whole (B_4). A bicompartmented glass cell with 1 mm liquid-thickness and spiral partition appears below (C). Filling hole-plugs are shown in center (Pl). To the right a glass drop-cell is shown with two dimpled recesses (F), and a gold-plated stainless-steel cell with drop-holder, taken apart to show the inner shell (G_1) and the outer shell (G_2). I is a photochemical glass cell with two concentric spaces for investigation of a light-absorbing reaction in the outer space, provided with photons from a light-producing reaction in the inner space. J is a cell for continuous rotation of biological systems with open-groove drop-holders at both end-discs (stainless steel).

FIG. 5. Vessel for rapid response. Drop-type vessel for rapid response, with 1 ml liquid-layer and 5 ml capacity. Drops are held in capillary spaces between arcuated baffles, obtained by indentation of the inner wall.

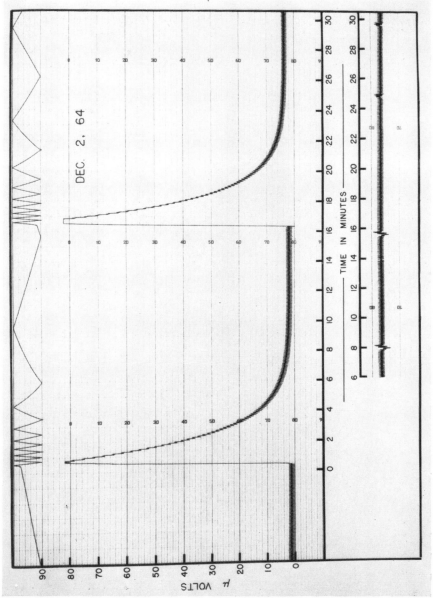

Fig. 6. Calibrations of pulse-microcalorimeter. A. Recording of heats of two consecutive reactions from a continuously rotating vessel. The two zero spikes recorded at 34 and 42 min pertain to the repeated tumbling motion by which the first reaction had been initiated, those taken at 51 and 56 min to the tumbling motion by which the reactant of the second reaction had been swept from the grooves in the disc-drop-holder.

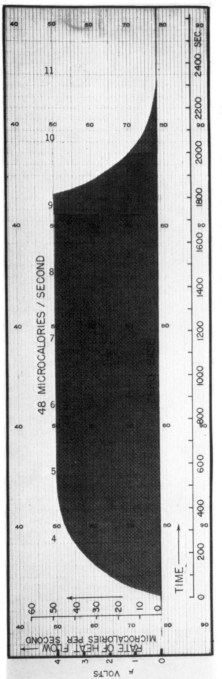

FIG. 6. Calibrations of pulse-microcalorimeter. B. Calibration of pulse-microcalorimeter with steady electric current.

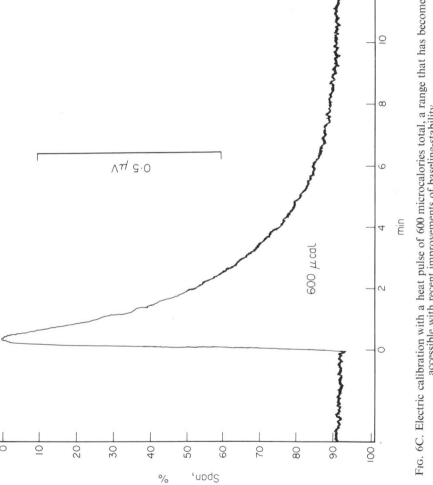

FIG. 6C. Electric calibration with a heat pulse of 600 microcalories total, a range that has become accessible with recent improvements of baseline-stability.

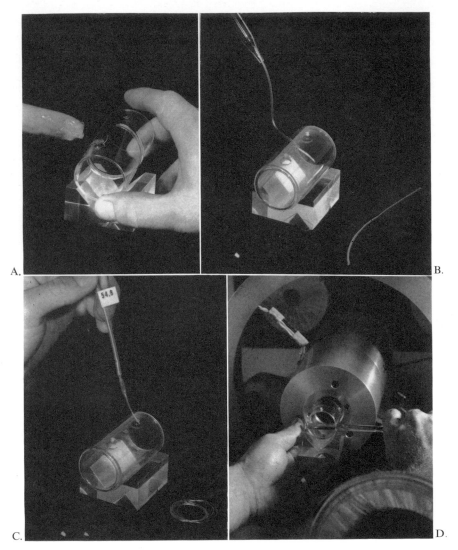

FIG. 7. Manipulation of reaction-vessels. A. Lubrication. B. Filling of main space.
C. Filling of drop-wells. D. Insertion of vessel in block.

in the 10 mM glutathione experiment (lower curve in Fig. 9A), the reaction did not go to completion, nor was it complete at 2290 cal in the test with 50 mM, because even more, 3740 cal, were observed at 100 m molarity. More data including higher concentrations and heats of dilution and activity coefficients would permit an extrapolation to ΔH, the total heat of reaction, and a thermodynamic analysis of this or similar systems, namely, the determination of the equilibrium constant, heat, free energy, and entropy change of the reaction. This example has been given to show that heat may be used to observe new reactions and to study their thermodynamics, even in cases where the product cannot be isolated and where there would be no other method of analysis for the product.

2. Enzymes and Substrates

With rare exceptions (where ΔH is zero), no chemical change is hidden from detection by calorimetry. In a mixture of any number of enzymes, each endividual enzyme will reveal its presence and its catalytic activity by the evolution or absorption of heat when the matching substrate is added in the calorimeter. Vice versa, in a mixture of any number of substrates each individual substrate will reveal its presence and quantity by heat changes when an enzyme specific for the substrate is added to the bulk mixture. Illustrations for this argument are the experiments of Fig. 9B. The objective was to find whether or not an active preparation of the enzyme glutaminase contained as a contaminant another enzyme attacking glutamic acid. A slow but measurable heat production in the upper recording revealed the presence of a contaminant, which could then be removed in two successive steps, with the results as shown by the middle and lower recordings. Higher amplification had to be used to uncover some still remaining activity of a lower order of magnitude, after the first purification. In the third recording, taken after a second purification procedure, no heat is visible, in spite of even higher amplification of the potential. The instantaneous, transient heat changes upon mixing were caused by dilution. They are readily distinguished as a side-phenomena from the continued enzymic heat production. Differences in kinetics frequently permit one to observe and to eliminate side-reactions.

This application of calorimetry may become useful when enzymes have to be assayed routinely concerning both activity and purity of the preparations. Any number of different enzymes may be investigated for both aspects with only one method and instrument, the calorimeter, supposing that pure preparations of the substrates under consideration are available for reference. On the other hand, the purity of substrates may be checked in the same manner with pure preparations of enzymes. This could mean simplification in comparison with the alternative, numerous other methods of analysis, each of which would be applicable to only one reaction, or one group of reactions.

3. The Hydrolysis of Adenosinetriphosphate and the Neutralization of Protons

For adenosinetriphosphate hydrolysis, the principal energy-donating reaction in the various work-performances of living matter, the reaction-heat had been estimated previously at $-12,000$ cal/mole, and designated as the heat of an "High-energy phosphate bond".

Pulse-calorimetry[6,7,12] revealed a much lower heat, -4800 cal/mole. The classical overestimate was due presumably to an accompanying heat of neutralization of one proton after the reaction in buffer.

$$(1) \quad ATP^{4-} + H_2O \rightarrow ADP^{3-} + HPO_4^{2-} + H^+$$
$$(2) \quad H^+ + BOH \rightarrow B^+ + H_2O$$

In this case, the side-reaction could not be identified by differences in kinetics. It is, however, recognized immediately when the reaction is carried out twice, in two buffers of different heats of neutralization (Table 1).

TABLE 1

	ΔH In trishydroxymethyl aminomethane buffer	In $PO_4^{\Delta H}$ buffer
Hydrolysis + neutralization	$-16,400$	-6800
Neutralization alone	$-11,600$	-1900
Hydrolysis	-4800	-4900

When properly treated, this phenomenon is analytically useful. In cases where a reaction is incompletely formulated, it permits one to detect whether or not and how many protons or hydroxyl ions are produced or disappear (see, for example, Laki and Kitzinger[13]).

4. Fibrinogen–Fibrin Conversion

When the blood protein, fibrin, interacts with the enzyme, thrombin, it polymerizes into a network of fibers. However, prior to the clotting, a splitting of peptide bonds takes place on the protein through proteolytic activity of thrombin. Analysis by ultrasensitive calorimetry revealed in a study by K. Laki and C. Kitzinger (1956)[13] that the number of peptide bonds that are split is two. In phosphate buffer with a heat of neutralization of approximately −1900 cal/mole the total heat evolved is −44,000 cal/mole. This heat contains three contributions:

(1) heat of splitting of peptide bonds;
(2) heat of clotting; and
(3) heat of neutralization of the protons liberated in the peptide splitting.

When the calorimetric experiment is repeated in "Tris"-buffer with a heat of neutralization of −11,600 cal/mole, the total heat evolved is approximately −64,000 cal/mole of protein, not 44,000. The difference of ∼20,000 cal/mole is due to a ∼10,000 cal/mole⁻¹ difference between the heats of neutralization of the two buffers. Two protons, or the splitting of two peptide bonds, no more, no less, are therefore involved. It is important to carry out the two experiments in the range of pH 7.5 to 9.0. At pH values below 7.0, protons appear to originate from histidine residues as a result of the polymerization and the total heat of reaction further increases.

5. Antigens and Antibodies

Antigen–antibody interactions are usually detected by insoluble compounds which they produce. Another method of analysis, not limited to any specific properties of the product, is calorimetry. In Fig. 9c, obtained by Steiner and Kitzinger,[14] a heat change of −3600 cal/mole was observed in the combination of human serum albumin as an antigen, and antibody-globuline of a rabbit immunized with human serum. With proper controls the authors established that no side-reactions were involved. In the lower recording the specific antibody was removed prior to the calorimetric test by precipitation with a limited quantity of the specific antigen. The supernatant did not appreciably interact with the antigen in the calorimeter. In separate controls human albumin was replaced by bovine or rabbit serum. Only the species-specific albumin gave the reaction producing the heat.

When the reaction between antigen and antibody is detectable in a specific way by the non-specific phenomenon of heat which it produces, the same technique might be successfully applied to other biologically or pharmacologically active compounds in search for their specific sites or partners in living systems. The quantities of substances applied in the experiments of Fig. 9c were small indeed in molar terms (0.94 μm), and so were the molarities, 1/20,000 м. This is particularly important when large molecules are involved as in the present example. It is also obvious that combustion of such compounds could never reveal the small difference in heat that exists between the reactants and products.

6. Changes in the Conformation of Polynucleotide Molecules

Ultrasensitive calorimetry permits the observance of changes in the conformation of large molecules. One case in point is the folding or stretching polynucleotide chains with alterations of ionic strength. At low ionic strength the chains are extended by the repulsive force of their charges. At high ionic strength, when the charges are neutralized, the molecules fold.

It would be impossible to observe the heat change during alterations of the ionic medium in the calorimeter. The large heat of dilution of the ionic medium with its high molarity would obscure the macromolecular change with its low molarity. However, the additivity of thermodynamic data permits an indirect measurement. An extended double helix may be formed from *extended* single strands of polyribouridylic and polyribo-adenylic acids in one experiment at low ionic strength, and from *folded* single strands in another experiment, at high ionic strength. The difference between these two heats is the heat absorbed upon stretching, or liberated upon folding the single strands (Steiner and Kitzinger[26]).

7. Surface Interactions

Heat changes associated with surface interactions were observed already during calibrations of the first heat-pulse instrument. With reasonably small quantities of substances in the calorimeter, heats of formation of water from hydrogen and hydroxyl ions, consistent with independent measurements obtained by others on large quantities, were observed. However, upon a downward extension into the sub-micromole range excess heat from a side-reaction became visible. From an unknown reaction, heat in the amount of 9200 cal/mole of chloride ions was observed when HCl was dropped into the NaOH solution of the main vessel space. Above $0.35 \, \mu M$ of Cl^- the total excess heat did not rise with increasing amounts of HCl added (Fig. 10). This meant that the unknown partner of the chloride ions in the heat-producing reaction was present in a limited quantity, $0.35 \, \mu M$. This amount would be sufficient by order of magnitude to form a monomolecular layer on the inner surface of the vessel. Proof of the assumption that the interface between solution and vessel was the unknown reaction-partner, came in two ways: (1) in teflon vessels the heat was only one-half the 9200 cal/nole observed with

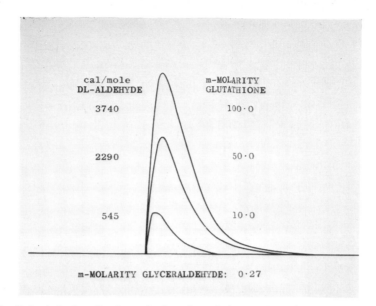

FIG. 9. Analytical applications of pulse-microcalorimetry. A. Unknown interaction between D,L-glyceraldehyde and glutathione and its reversibility demonstrated by calorimetry.
Explanations in text.

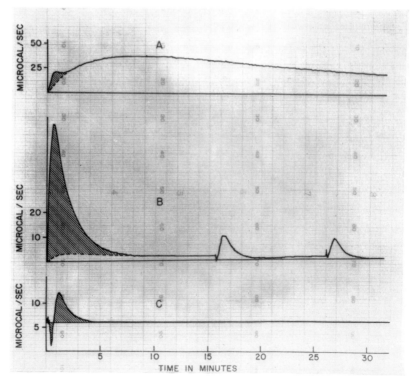

FIG. 9. Analytical applications of pulse-microcalorimetry. B. Calorimetric detection and assay of enzymes or substrates.
Explanations in text.

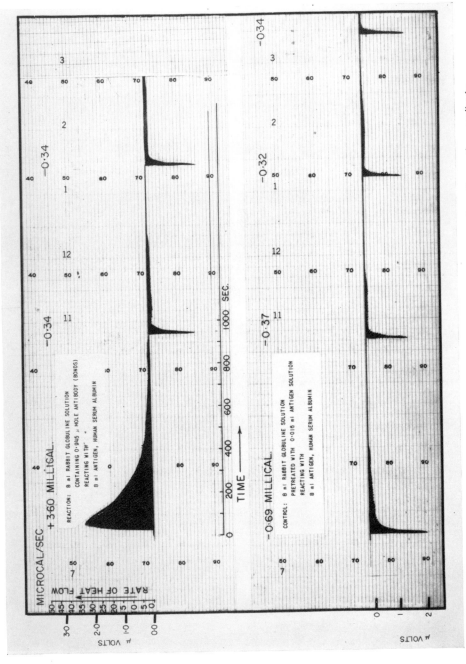

Fig. 9. Analytical applications of pulse-microcalorimetry. C. Demonstration of an antigen-antibody interaction by heat. Explanations in text.

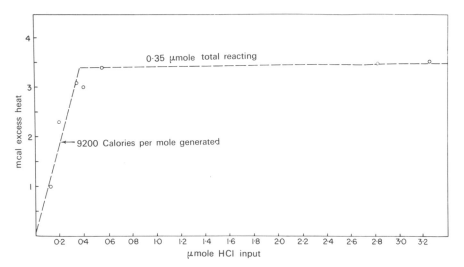

FIG. 10. Interaction of chloride-ions with walls of reaction-vessel (glass). The heat of reaction is 9200 cal/mole. The walls are saturated when 0.35 μM of chloride-ions form a monomolecular layer at the solid–liquid interface.

glass vessels; (2) addition of chloride (m/10,000 NaCl) to the NaOH in the main vessel space (before addition of the acid) eliminated the side-reaction. The simple expedient of changing the amounts of the substance had uncovered and clarified in a quantitative manner a reaction of which the chemist is not usually aware—that the dissolved objects of his work interact with the walls of the containers.

In many classical calorimetric experiments of the past the same phenomenon and small errors from the same source must have been present, and remained unnoticed. With a smaller surface-to-volume ratio in classical calorimeters the error was smaller than it is in pulse-calorimetry with the large surface-to-volume ratio of the special vessels. For the new technique the elimination of this source of error by addition of ions was, therefore, particularly important. The fact that the phenomenon could be observed at all indicated a considerable power of the method in treating phenomena at interfaces between solids and liquids. Such problems are encountered, for example, in catalysis, in membrane transport, in macromolecular biology and in all kinds of chromatographic methods for chemical separation.

8. Life-detection by Direct Calorimetry

Many different analytical methods are under consideration for attempts to discover living organisms on other planets. Heat offers the advantage of being the most universal index of metabolic activity. Calorimetry may therefore become a useful tool in the endeavor of life-detection.

A preliminary study has been made by T. B. Weber[16] on samples of desert soil (200 mg) as well as with cultured microorganisms. Typical

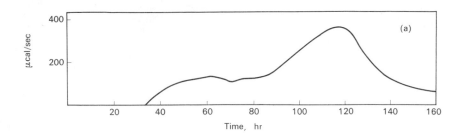

FIG. 11. Detection of living matter
A. Thermogram of a desert soil sample in a synthetic medium containing glucose, lactate and the ions K^+, Na^+, SO_4^{2-} and PHO_4^{2-} (T. B. Weber).

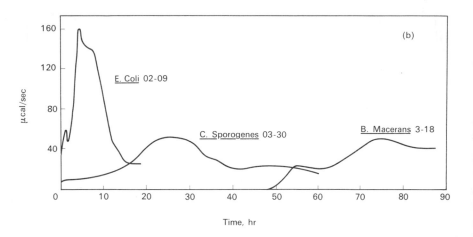

FIG. 11. Detection of living matter
B. Thermograms obtained with small inocula of three different microorganisms in organic culture-medium.

heat curves were obtained, even after long periods of dormancy, as appears from Figs. 11A and B. In the soil sample (Fig. 11A), a growing metabolic rate was observed after a dormant period of 36 hr. Inoculation of organic culture-media with small numbers of various microorganisms also resulted in typical heat- or growth-curves (Fig. 11B), in one example after a latent period of almost 50 hr. On their experiments with *E. coli*, on the other hand, the authors report: "Using the given inoculum, the microcalorimeter was able to detect growth in about four minutes from the time of inoculation, while by plate-count, gravimetry and photometry an elapsed time of two to three hours was required." It appears from the study that the heat observed was mostly associated with *growth*, not *basal* metabolism of the microorganisms. Approximately 10^5 to 10^6 cells, or a minimum mass of 1 μg, are given as the limit of detectivity with the heatburst-microcalorimeter (applied in this instance to steady, not burst-like production of heat).

D. THERMODYNAMIC MEASUREMENTS BY ULTRASENSITIVE CALORIMETRY

Thermodynamic analysis of any process includes two measurements, the heat change, ΔH, and the entropy term, $T\Delta S$. Their sum, the free energy change, ΔF, obtainable from chemical equilibria, requires for complete thermodynamic understanding an additional measurement of ΔH, from which the entropy term, $T\Delta S$, follows. For direct determinations of ΔH the role of ultrasensitive calorimetry is obvious. Less obvious or expected was its potential for the determination of ΔF and ΔS from purely thermal data, without chemical analysis, by means of only two measurements of heat.[17]

While third-law entropy determinations are carried out with measurements of heat capacity on reactants and on products, ultrasensitive reaction calorimetry determines entropy by measurement of heat during the reversible conversion of reactants into products.

The argument on which the method is based is very elementary. When a reversible reaction is permitted to proceed in the calorimeter in dilute solution, first with the reactants and then in a second experiment with the products as the starting material, heat changes of opposite sign are observed (Fig. 12). The heat change, Q_1 [cal/mole^{-1}], of the first experiment is proportional to the quantity of products formed, which is $Q_1/\Delta H$, where ΔH is the unknown heat of the complete transformation. The heat change, Q_2 [cal/mole^{-1}] of the second experiment is proportional to the quantity of reactants formed from products, which is $-Q_2/\Delta H$. Therefore, the expressions $Q_1/\Delta H$ and $-Q_2/\Delta H$ may be substituted for the amounts — or concentrations — of reactants and products, present at the end of either of the two experiments, which is the state of equilibrium. In the given example of a reaction with two equimolar participants on either side, the squared ratio of activities, calculated from the concentrations found by heat, is the equilibrium constant, K. From K the free energy change is readily derived. It is $\Delta F° = -RT \ln K$ (Fig. 12). The heat change, ΔH, is $Q_1 - Q_2$, because in the second reaction, with its sign reversed,

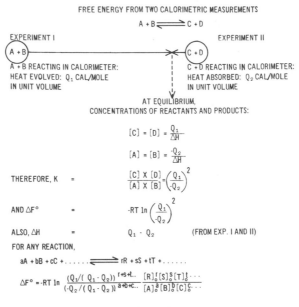

FREE ENERGY FROM TWO CALORIMETRIC MEASUREMENTS

$$A + B \rightleftharpoons C + D$$

EXPERIMENT I   EXPERIMENT II

(A + B)   ⟩⟨   (C + D)

A + B REACTING IN CALORIMETER: C + D REACTING IN CALORIMETER:
HEAT EVOLVED: Q_1 CAL/MOLE   HEAT ABSORBED: Q_2 CAL/MOLE
IN UNIT VOLUME   IN UNIT VOLUME

AT EQUILIBRIUM,
CONCENTRATIONS OF REACTANTS AND PRODUCTS:

$$[C] = [D] = \frac{Q_1}{\Delta H}$$

$$[A] = [B] = \frac{-Q_2}{\Delta H}$$

THEREFORE, K $\quad = \quad \dfrac{[C] \times [D]}{[A] \times [B]} = \left(\dfrac{Q_1}{-Q_2}\right)^2$

AND $\Delta F° \quad = \quad -RT \ln \left(\dfrac{Q_1}{-Q_2}\right)^2$

ALSO, $\Delta H \quad = \quad Q_1 - Q_2 \qquad$ (FROM EXP. I AND II)

FOR ANY REACTION,

$$aA + bB + cC + \ldots \ldots \rightleftharpoons rR + sS + tT + \ldots \ldots$$

$$\Delta F° = -RT \ln \frac{(Q_1/(Q_1 - Q_2))^{r+s+t\ldots} \; [R]_o^r [S]_o^s [T]_o^t \cdots}{(-Q_2/(Q_1 - Q_2))^{a+b+c\ldots} \; [A]_o^a [B]_o^b [C]_o^c \cdots}$$

FIG. 12

the incomplete first reaction that began with the pure reactants is continued to completion, where only products are present.

Compared with third-law entropy determinations this approach has the advantage of simplicity. Instead of a summation of an infinite number of infinitesimal steps, it requires only two calorimetric measurements for the determination of both ΔS and ΔH. Instead of requiring the techniques of low-temperature physics, it can be carried out at room temperature. Instead of requiring enormous quantities (order: 10 g), it works with small amounts. Instead of demanding absolute purity, it demands only purity from interacting contaminants. It is not subject to errors from freezing of non-ideal structures. It is not subject to uncertainties from extrapolations to the unreachable absolute zero of temperature. It does not necessitate corrections for entropies of solution. It does not rely for the determination of entropies of reaction, on the subtraction of two very large quantities – entropies of reactants and entropies of products – to observe a small difference between the two with sufficient accuracy. And finally, in connection with the subject of this book, it does not exclude measurements on proteins that might be altered irreversibly during an approach to absolute zero.

However, the new procedure depends, like the classical chemical methods for the same task, on the relation $\Delta H - T\Delta S = -RT \ln K$. It is feasible only when $\Delta H - T\Delta S$ is not too large, only when the reverse reaction proceeds to an extent that can be measured. The limit depends upon the sensitivity of the calorimetric instrumentation. It is at present 3500–4000 cal/mole for $\Delta H - T\Delta S$. Distinct from its classical precursor—the combination of a calorimetric measurement for ΔH with a *chemical* determination of equilibrium and ΔF—the purely calorimetric method for entropy-determination does not borrow from the non-thermal techniques of inorganic, organic, and biological chemistry. It will be chosen, particularly, when other methods do not permit one to detect and to measure small quantities of reactants in the presence of enormous quantities of products at equilibrium.

The applications of ultrasensitive calorimetry to thermodynamics will now be shown with concrete examples, first for the measurement of heat of reaction, ΔH.

1. Heat of Reaction: Asparagine-hydrolysis

Classical combustion heats, measured on crystalline substances, must be corrected for heats of solution, dilution and, sometimes, ionization, to represent heats of reaction in aqueous solutions. When heats of reaction are measured directly, by reaction-microcalorimetry, these procedures are greatly simplified. Nevertheless, one must carefully observe a standard procedure soon to be described, for the following reasons. No interaction between chemical individuals in separate solutions can be initiated without mixing and thereby diluting both participants. This introduces disturbing heats of dilution. Moreover, dilution also occurs upon mixing of the accessory media such as buffers or salt solutions employed to maintain pH and ionic strength at desirable levels. Furthermore, when for various reasons partner-solutions of different pH are mixed, ionizations take place. These, too, manifest themselves by liberation or absorption of heat. It is important to avoid errors from these sources with all certainty. Individual determinations and subtraction of the various heats of dilution and ionization would not suffice. Some of the accessory heat changes must not be applied as corrections because they are reversed when the species with whom they are associated disappear in the reaction.

A simple and thermodynamically correct determination of ΔH is possible, nevertheless, with a certain standard procedure. To demonstrate this procedure, we choose as an example the hydrolysis of a biological amine, asparagine, into aspartic acid and ammonia.

$$\text{Asparagine} + \text{water} \rightarrow \text{aspartate}^{+--} + \text{NH}_4^+$$

The objective is to measure the heat of this transformation with reactants and products in standardized states, free of accompanying heats of dilution or ionization.

In a first experiment, using drop-type calorimeter vessels, droplets of pure reactant in a standardized state (aqueous, 0.1 molar solution of asparagine at pH = 7.0) are added to a bulk of buffered enzyme solution (enzyme asparaginase, in 0.01 M borate buffer, pH = 8.5). The total calories liberated in this reaction are measured as shown in Fig. 13. This heat (−5140 cal/mole) was associated with the transformation of the reactant, asparagine, in a standardized state (aqueous, 0.1 M, pH = 7.0) into products of a less well-defined state (aspartic acid ions and ammonia

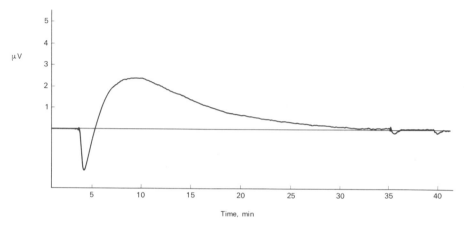

Fig. 13. ΔH-determination, first step. To eliminate corrections for heats of dilution, ionization, surface interactions and reverse reactions, determinations of ΔH are carried out in two steps. In the example shown here asparagine 0.1 M, aqueous, pH 7.0 is hydrolyzed, diluted and ionized in a buffered enzyme solution. The heat change is −5140 cal/mole. To obtain the correct value of ΔH a second experiment, Fig. 14, is required.

ions, aqueous, more dilute, in the buffered enzyme-solution, pH = 8.5). Therefore, a second experiment is required after which the poorly defined products should emerge standardized (0.1 M, pH = 7.0) from their dilute solution at different pH (8.5) in the buffered enzyme mixture. While it is obviously impossible to "remove" standardized products from the enzyme solution for measurement of the accompanying heat change, it is easy to perform in a calorimeter the reverse reaction. Instead of being *removed* from the final solution after the first experiment, the products, in droplets of standardized state (0.1 M, pH = 7.0) are *added* to a quantity of enzyme buffer solution identical to the one used in the first experiment, in a second, a control experiment (Fig. 14). The heat change (an absorption of +570 cal/mole as shown in Fig. 14) is measured calorimetrically, and its sign is reversed. In adding this second measurement of heat to the first, we have observed the true heat change ($\Delta H = -5700$ cal/mole^{-1}) of asparagine hydrolysis (aqueous, 0.1 M, pH 7.0) in two successive steps:

(1) formation of certain unknown but defined species from standardized *reactants* (aqueous, 0.1 M, pH 7.0) and (2) formation of standardized *products* (aqueous, 0.1 M, pH 7.0) from the *same* unknown but defined species which are present after the first as well as after the second experiment.

When identical vessels and solutions are applied in the paired experiments, the procedure appears to eliminate also, whatever interactions may take place between vessels and solutions. It would eliminate heats of dissociation, occurring with the dilution of dissociable components in a buffer enzyme system. It would eliminate heats from interactions between

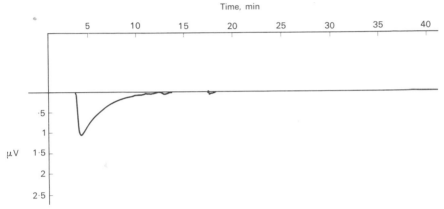

FIG. 14. ΔH-determination, second step. Ammonium and aspartate ions, aqueous, 0.1 M, pH 7.0, were "removed" as it were, from the solution of hydrolyzed, diluted and ionized products of the experiment Fig. 13, by dropping the ions into another sample of the buffered enzyme solution and reversing the sign of the heat change, $+570$ cal/mole. ΔH is the sum of the two heats observed in Figs. 13 and 14.

water as a solvent, and complex molecules or structures on which water may be bound, or altered in its molecular configuration. Likewise, the procedure as outlined would prevent, that errors in the measurement of ΔH are encountered when the reaction under study is measurably reversible and therefore incomplete, as will be shown in the following chapter. The procedure is therefore recommended for general use* in determinations of heats of reaction.

*While this procedure rules out reversibility of the reaction as a source of error it requires that the reaction reaches equilibrium or, in the case of practically irreversible reactions, completion. Only then the final compositions of the mixtures, after the first and after the second experiment, are identical as they must be. In enzyme systems with high Michaelis-constants, incomplete reaction as a source of error must be carefully considered. In the present example it was ruled out by high enzyme and low substrate-concentration, and verified by variation of the substrate-concentrations over a wide range. Figure 9 is another example of a complete transformation. Reaction-rates are a linear function of the distance from equilibrium. This shows that infinitesimal deviations from the end-point resulted in reactions at full rate, and that the reaction proceeded from both sides to the equilibrium-point.

2. Free Energy and Entropy of a Readily Reversible Reaction: Fumarase

For a readily reversible process, the fumarase-reaction of the Krebs-cycle, the heat, free energy and entropy changes were obtained (Kitzinger and Hems [18]) with the heatburst microcalorimeter on very dilute solutions. Two experiments (Fig. 15) were carried out at standard temperature, 298°K, one with the reactant, fumarate, the other with the product, malate. In this particular instance the evaluation is very direct and simple. On both sides of the reaction equation

$$\text{Fumarate} + H_2O \rightleftarrows \text{Malate}$$

there is (except for water which is taken as unity) only one single reactant or product. Therefore, the equilibrium constant is

$$K = \frac{[\text{Malate}]}{[\text{Fumarate}]}$$

or

$$K = \frac{Q_1}{-Q_2}$$

while the heat of reaction, ΔH, is $Q_1 - Q_2$. With $Q_1 = -3020$ [cal/mole] and $Q_2 = +680$ [cal/mole^{-1}] the equilibrium constant is

$$K = 4.45.$$

The combined driving force is

$$\Delta H - T\Delta S = -RT \ln K = -880 \text{ cal/mole}^{-1}.$$

The reaction is readily reversible; the heat of reaction is

$$Q_1 - Q_2 = -3700 \text{ [cal/mole}^{-1}].$$

The entropy term,

$$T\Delta S = -2820,$$

is of the same order of magnitude. It opposes the heat-term $\Delta H = -3700$ as a driving force, which is the reason for the reversibility of the fumarase-reaction. The standard entropy change of the reaction is:

$\Delta S = -9.46$ "entropy units" [cal/mole^{-1}/deg^{-1}], at temperature $T = 298°K$.

The data thus obtained by heatburst-microcalorimetry have confirmed the earlier classical analysis of the same problem with chemical methods by H. Krebs.[19] Here the question arises whether the convenient calo-

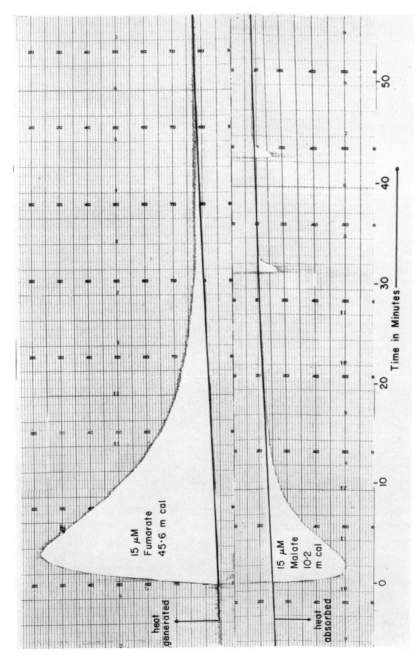

FIG. 15. Fumarase reaction, heat and free energy. Thermodynamic analysis by microcalorimetry, of the enzymic interconversion of fumarate and malate, a Krebs cycle reaction. The upper recording shows the heat liberated during conversion of a solution of fumarate into a solution containing fumarate and malate in equilibrium. The lower recording shows the heat absorbed when a solution of malate is transformed into the same equilibrated mixture of malate and fumarate. The evaluation of free energy, heat and entropy changes is shown in the text.

rimetric method of analysis may also be applied to reactions that are less readily reversible than the fumarate–malate interconversion.

3. Free Energy and Entropy of a Practically Irreversible Reaction: Glutaminase

Glutamine hydrolysis with the enzyme glutaminase

$$\text{Glutamine} + H_2O \rightarrow \text{Glutamate}^{+--} + NH_4^+$$

was considered an irreversible reaction for all practical purposes. Nevertheless, the heat pulse microcalorimeter demonstrated its reversibility[20] (Fig. 16). When concentrated enzyme was added to a 0.88 molar solution of glutamic acid and ammonia, heat was absorbed in synthesis of glutamine at an initial rate of ~10^{-8} moles/sec.

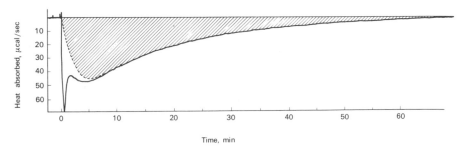

FIG. 16. Reversibility of the glutaminase reaction. Reversibility of the glutaminase reaction is demonstrated by calorimetry. The heat absorption is due to enzymic synthesis of glutamine from glutamic acid and ammonia, 0.884 M. Net heat absorption declines (recording rises to the zero base) because of simultaneous rehydrolyzation of the newly formed glutamine.

To sustain such a low rate of glutamine-synthesis, an almost inexhaustible amount of glutamic acid and ammonia (1.3×10^{-2} moles) was present in the calorimetric vessel. Why, then, was the initial rate of heat absorption not sustained, why did the negative deflection of the calorimeter-potential rise to the zero base? The reason is, that simultaneously with heat *absorption* in glutamine synthesis, a mounting *liberation* of heat takes place as newly synthesized glutamine is *rehydrolyzed* at a rate that increases with its rising concentration. Finally, together with the *rising* glutamine concentration the rate of hydrolysis levels out. It approaches the *constant* rate of glutamine-synthesis. The net heat output becomes zero, although glutamine synthesis continues to proceed at its initial, undiminished rate of ~10^{-8} moles/sec. Chemical equilibrium is attained. The forward and reverse rates of reaction are identical.

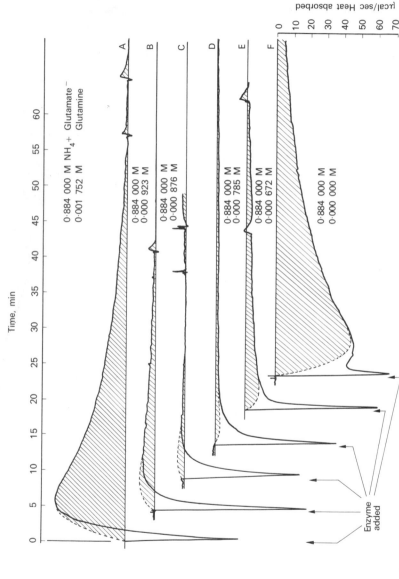

Fig. 17. Equilibrium of the glutaminase reaction. Six recordings in the calorimeter demonstrate the meaning of chemical equilibrium. Zero heat would have been observed with a solution of glutamine and ammonium glutamate in chemical equilibrium. The mixtures tested show either absorption of heat (when the assay contains *more* glutamine) or liberation of heat (when it contains *less* glutamine than required for equilibrium). The equilibrium-point is found by linear intra- or extrapolation (see Fig. 10).

Glutamine is now present at an unknown concentration in equilibrium with 0.88 M glutamic acid and ammonia. This unknown concentration and with it the equilibrium constant, our objective, are found by linear extrapolation between close estimates. In Fig. 17, six experiments are shown, in which small amounts of glutamine had been added prior to calorimetry, to the 0.88 M solution of glutamic acid and ammonia. In some of these experiments, net heat was absorbed; in the others, liberated. If in one further test exactly the right amount of glutamine would have been added, reactants and products would have been at equilibrium to begin with. No heat at all would have been observed. In Fig. 18, the total amounts of heats observed were plotted against the amounts of glutamine added. The

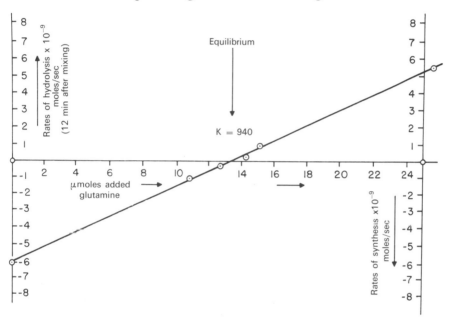

FIG. 18. Equilibrium-constant by interpolation. Net reaction rates (or total heats measured) in the experiments of Fig. 9 permit the finding of the point of equilibrium by interpolation. Linearity is caused by linear relation between hydrolysis-rate and glutamine-concentration. A constant rate of glutamine synthesis causes the displacement of the intersect from zero to -6×10^{-9} moles/sec.

intrapolated point of zero heat, that is the point of chemical equilibrium, was observed at a ratio of 940 of reactant over product-concentrations. This ratio is K_c, the apparent equilibrium constant. Correction of K_c by activity coefficients leads to the thermodynamic equilibrium constant, K.

The linear characteristics of the plot, Fig. 18, are significant for several reasons:

1. Any pair of two of the six experiments would have permitted the

determination of the equilibrium constant and free energy. No more than two experiments are required for this method.

2. No "dead-zone" was reached as equilibrium was approached from both sides. Infinitesimal deviations from equilibrium resulted in net heat production or absorption, in net hydrolysis or synthesis of glutamine, at identical rates. The rate of synthesis was constant throughout, and identical in the six experiments, at equilibrium and right or left from the equilibrium. Only the rate of hydrolysis varied. It was related linearly to the amounts of glutamine added, to the low concentrations of glutamine which varied from zero to 1.57 millimolarity.

3. The recorded graphs and the linear plot demonstrate the nature of chemical equilibrium. With widely disproportionate mass-actions, the forward reaction-rate becomes identical to the reverse reaction-rate at a precisely determined point.

The sensitivity of the method can be further exploited as shown in Figs. 19 A and B. In two and four-times less concentrated solutions of glutamic acid and ammonia, equilibria could still be determined.* However, the limit of the method is likely to be reached in a reaction of this type before the free energy exceeds -4000 cal/mole^{-1}, with presently available instrumentation.

Reference is made to the original publications[20,21] concerning corrections and the final data for the glutaminase reaction (pH 7.0, $T = 25°C$):

$$\Delta H° = -5160 \text{ cal/mole}^{-1}$$

$$\Delta F° = -3420$$

$$T\Delta S° = -1740$$

$$\Delta S° = -6.0 \text{ cal/mole}^{-1}/\text{deg}^{-1}$$

As in the fumarase-reaction, so in the glutaminase-reaction, the heat and entropy terms are opposed as driving forces.

It remains to be discussed, why in reactions that are almost irreversible, very small transformations of energy can be used with confidence for a determination of free energy. Could these small heats not have been, at least in part, a product of misleading side-reactions? They could not, for only a reaction and its own mirror-image have strictly identical kinetics (except for reversal of their signs) at either side of the equilibrium, where mass-actions are appropriately shifted. If the kinetics were not otherwise identical, the heats absorbed and produced could not possibly cancel as they do at the equilibrium point. In our example (Figs. 16, 17

---

*The apparent equilibrium constants are different in Figs. 9, 11 and 13, only because of the differences in activity. When the thermodynamic equilibrium constants were calculated with known activity coefficients, they agreed within $\pm 3\%$ under these different conditions.[8] Because of the logarithmic relation, the error in ΔF is extremely small.

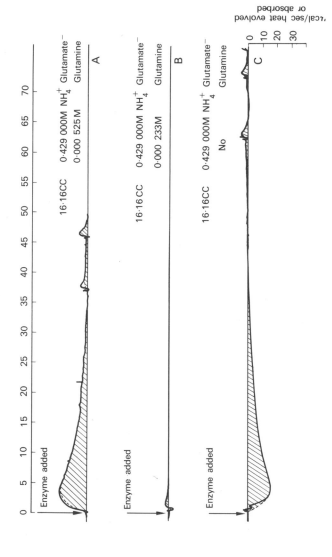

Fig. 19A. Glutaminase equilibrium in 0.43 M glutamate. Glutaminase equilibrium demonstrated with a nearly perfect ratio of reactant and product-concentrations (middle recordings) with heat liberated (above) or absorbed (below) upon small deviations from equilibrium. Absence of side-reactions is proven by the identical kinetics of the heat curves.

137

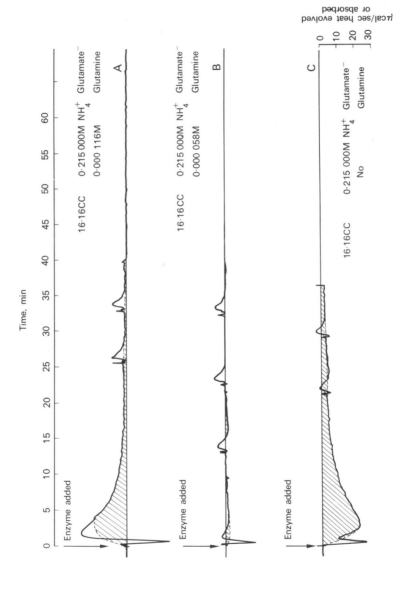

FIG. 19B Glutaminase equilibrium in 0.21 M glutamate. Glutaminase equilibrium at lower concentrations of reactants and products. For explanation see Figs. 9 and 11.

138

and 19) the reaction that absorbed the heat is, therefore, safely identified as the enzymic synthesis of glutamine, not previously observed without ATP as an energy donor.

4. Thermodynamics of Adenosinetriphosphate Hydrolysis

The thermodynamic analysis of the glutaminase reaction as shown was selected as an example for a further reason. It has led indirectly to finding the thermodynamic quantities of adenosinetriphosphate hydrolysis (the main energy-donating reaction for the various functions of living matter). This was possible by means of two enzymically catalyzed reactions (1) and (2).

(1) Glutamine $+ H_2O$ $\rightarrow$ Glutamate^{+--} $+ NH_4^+$
(2) Glutamate$^{+--} + NH_4^+ + ATP^{4-} \rightarrow$ Glutamine $+ ADP^{3-} + HPO_4^{2-} + H^+$
(3) $H_2O + ATP^{4-} \rightarrow$ $ADP^{3-} + HPO_4^{2-} + H^+$

The free energy of ATP hydrolysis, reaction (3), is negative and large, with an equilibrium too one-sided for a direct measurement. However, reaction (3) may be considered as the sum of two coupled steps, reactions (1) and (2), the equilibria of which are measurable.[21,22]

ΔH, the heat of ATP hydrolysis, was directly determined with myosin as the enzyme, in experiments as shown in Fig. 20.[10] The data finally obtained at 25°C in the presence of Mg^{++} ions were:

$\Delta H = -4800 \text{ cal/mole}^{-1}$ $T\Delta S' = +2200 \text{ cal/mole}^{-1}$
$\Delta F' = -7000 \text{ (pH 7.0) cal/mole}^{-1}$ $\Delta S' = +7.5 \text{ cal/mole}^{-1} \text{ deg}^{-1}$

5. Free Energy and Heat of Biochemical Reactions

Adenosinetriphosphate hydrolysis, the fumarase reaction, and the glutaminase reaction are typical and widely different examples for a discussion of the relations between heat and free energy in biochemical systems.

Classical human and animal calorimetry had accepted heat as the expression of metabolic activity, without consideration of entropic contributions. Such an approach, while useful in whole-animal research, fails when the individual steps of the metabolic process are under consideration. The magnitude of entropic contributions can make the free energy either much smaller, or much larger than the heat of reaction, as shown in Table 2.

TABLE 2

	Heat	Standard ΔF	Physiological ΔF
Fumarase-reaction	−3700	− 880	− 880
Glutaminase-reaction	−5160	−3420	− 7620†
ATP-hydrolysis	−4800	−7000*	− 12,200‡

*ΔF^1 at pH = 7.0 in the presence of Mg^{++}.
†One-millimolar concentration of participants.
‡pH = 7.4 and 1-millimolar concentration of participants.

For these discrepancies between heat and free energy there are various reasons:

1. Reactant and product-molecules often have substantially different entropies under standard conditions. These make the free energy either larger or smaller than the heat. (Standard free energy is less negative than

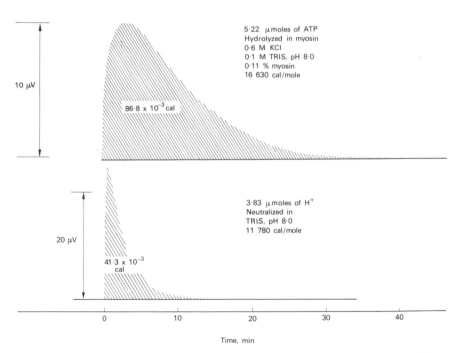

FIG. 20. ATP-hydrolysis and proton-neutralization. Enzymic hydrolysis of adenosinetriphosphate with myosin in the calorimeter. The heat of ATP-hydrolysis includes the heat of neutralization of one proton in the buffer. The heat of ionization of trishydroxymethylaminomethane was separately determined (lower recording) and subtracted. The difference, − 4800 cal/mole, is the heat of ATP-hydrolysis proper.

heat by 2820 cal/mole in the fumarase example, and by 1740 cal/mole in the glutaminase example.)

2. The free energy depends upon concentration or dilution in reactions where the number of reactants differs from the number of products. In the fumarase reaction, the number of participants on either side of the reaction equation is one. Therefore, the equilibrium does not shift upon dilution. When the reaction-mixture is diluted by a factor of 10, the free energies of both, the reactant and the product, decrease at temperature $T = 310°K$ by 1400 cal/mole each. Therefore, the difference between the two free energies, the free energy of reaction, is the same, regardless of whether the reaction-mixtures are concentrated or dilute.

In contrast, for the glutaminase reaction, the number of reactants is one, the number of products, two. Tenfold dilution at $K = 298°K$ changes the free energy of the products by -2800 cal/mole^{-1}, the free energy of the reactants by -1400 cal/mole^{-1}. Therefore, the free energy of the reaction increases by -1400 cal/mole^{-1} with ten-fold dilution and by -4200 cal with 1000-fold dilution (1 millimolar concentration is a reasonable assumption for concentrations of the participants in living tissue).

3. Even more powerful is the influence of pH upon the free energy of reactions in which protons are liberated or absorbed. ATP-hydrolysis belongs into this class. When carried out in one-molar, fully dissociated acid, the reaction

$$ATP^{4-} + H_2O \rightarrow ADP^{3-} + HPO_4^{2-} + H^+$$

would be *endergonic*. At pH $= 7.0$ and unit molarity of the other participants, the free energy is $\Delta F^1 = -7000$ cal/mole^{-1}. It includes the entropy and free energy of *dilution* of protons from unit molarity (inside the ATP molecule) to 10^{-7} molarity in neutral, buffered solution after hydrolysis, a contribution of $7 \times -1400 = -9800$ cal/mole^{-1}. On the other hand, the free energy of the subsequent *neutralization* of the protons in the buffer at its equilibrium point is *zero*, while the heat of this neutralization can be very large.

In tissue, the dilution of the participants and the slightly alkaline reaction add further to the free energy of ATP hydrolysis. At $T = 298°K$, pH $= 7.4$ and 1-millimolar concentration it is $\Delta F = -12,200$ cal/mole^{-1}.

4. In tissues, ATP as well as ADP form complexes with magnesium ions. The free energies of formation of the complexes are different for the two compounds. As a result, the free energy is 700 cal/mole more negative in the unphysiological condition, absence of Mg^{++} ions. The difference is merely additive. It is independent of pH and ATP concentration.

The complexity of the relations between heat, free energy and conditions in tissues as described has met with varied reactions during the short history of thermodynamic considerations of living systems. At the earliest stage, heat and free energy were considered nearly equal. Entropy differences were simply ignored. At a later stage, although the importance of entropies of reaction was recognized, the high free energy of ATP-hydrolysis was yet attributed to a "high-energy phosphate bond". From the heat measurement, $\Delta H = -4800$ cal/mole^{-1} [10] and from the considerations with Table 2, it follows that it is mainly the free energy of proton-dilution which supports the heat term as a driving force to such an extent that ATP can serve as the principal energy donor of almost every performance of living systems.

Recently, some investigators concluded from the complexity of factors that thermodynamic data are of limited value in biochemical research. Nothing could be more discouragingly incorrect. Thermodynamic data remain the basis of all causal thought concerning the nature of life. Reasonable assumptions on concentrations can be, and need to be made. If local structures cause unusual local concentrations of reactants or

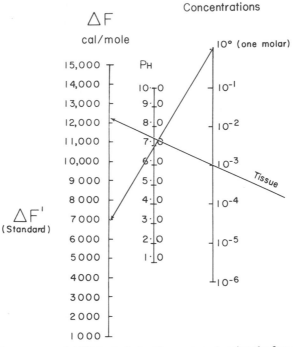

Fig. 21. Free energy of ATP-hydrolysis. Nomogram showing the free energy of ATP-hydrolysis as a function of pH and the initial concentration of ATP. By drawing a straight line through pH and ATP-concentration, the free energy is obtained at left, for ATP-hydrolysis in the presence of Mg^{++}. If Mg^{++} is not present, the free energies are more negative by -700 cal/mole^{-1}. Two examples are shown, the standard value ΔF and one typical of experiments with living systems.

products, these are still surrounded with solutions from which, and into which, the participants flow by diffusion. To these solutions the laws of thermodynamics apply without restriction, and regardless of intermediary steps at solid–liquid interfaces.

It is, therefore, essential not to surrender in view of the complexity of factors influencing the equilibria and free energies of chemical change. It is important to know the rather simple rules of treatment. The graph, Fig. 21, permits one to find for any given pH and concentration of reaction-participants the free energy of ATP-hydrolysis in the presence or absence of Mg^{++} ions. Results obtained with these considerations are consistent with a high, but not unreasonably high, efficiency of bio-chemical systems. One example may be quoted:

In the oxidative phosphorylation of mammalian mitochondria and also in the oxidative phosphorylation of chemosynthesis (L. Kiesow[23]), six molecules of ATP are synthesized with the free energy of oxidation of two molecules of reduced pyridine-nucleotide by oxygen as the terminal electron acceptor. In Kiesow's experiments, the conditions were those assumed in Table 2: $\Delta F = +73{,}200$ cal/mole^{-1} for 6 moles of ATP formed. The driving energy of DPN·H-oxidation is $\Delta F = 108{,}000$ cal/mole^{-1}, the efficiency, therefore, 67.5%, a value consistent with the assumption that the efficiency of living systems is neither poor, nor perfect.

6. Thermodynamics of an Antigen–Antibody Interaction: Human Serum Albumin Against Rabbit Antibody

Certain immunoreactions are reversible to such a degree that ultra-sensitive chemical methods of reactant-identification have permitted the determination of equilibrium constants. Singer and Campbell[24] found -7500 cal per bond for the free energy of interaction of bovine serum albumin, the antigen, with bivalent serum antibody-globulins produced in a rabbit. However, in attempting to estimate the heat of reaction, the authors did not find, within their experimental errors, a measurable effect of temperature upon the equilibrium. This finding excluded a large heat of reaction and strong bonding-energies between the antibody and the antigen. However, the measurement of the free energy did not exclude an endothermic nature of the process. It did not permit a safe estimate of the entropy change. Ultrasensitive calorimetry revealed the thermo-dynamic data in a study by R. Steiner and C. Kitzinger.[14]

The reaction between human serum albumin and the specific serum antibodies of the immunized rabbit was found not to be endothermic, nor is the heat evolved negligible as a binding force between the two pro-tein-molecules. It is substantial: -3600 ± 1000 cal per bond of the bivalent antibody. This measurement of ΔH permitted a fair estimate of the entropy difference between reactants and products: $+13$ entropy units [cal/mole^{-1}/deg^{-1}]. The entropy term $-T\Delta S$ at tissue temperature is approximately -4000 cal/mole^{-1}. The positive entropy change might be attributed to the release of bound, or polarized, water from both bonding sites. However, there may be contributions from configurational changes of either one or both proteins upon association, or from a rearrangement of the ionic atmospheres which surround them.

The two driving forces of the reaction have similar magnitude. Their concerted action represents a considerable free energy, -7500 cal/mole^{-1}. This order of magnitude of forces is familiar from other practically irreversible events in biochemical systems. In the present instance, as in ATP-hydrolysis the entropy term supports the heat term as a driving force, and makes the reaction virtually irreversible. The biological result is the elimination of a noxious agent.

This study of antigen–antibody interaction is remarkable also from another aspect. The exclusion of misleading side-reactions becomes increasingly difficult and important with low heat and with low molarities of solutions of compounds with high molecular weight. The multiplicity of proteins and other factors present in biological fluids and tissues and the difficulties of analysis and separation call for stringent and multiple control experiments. These demands have been satisfactorily met, as appears from the original paper.[18] The reasoning and the procedures applied may well become standard in future efforts toward similar objectives.

The modest heat of reaction and the enormous size of the molecules made these measurements difficult. Solutions were approximately 1/20,000 molar. Total amounts of antibody in the calorimeter ranged from 1 to 0.2 μM, the heats evolved, from 4.6 to 1 mcal total. The sensitivity of the heatburst-microcalorimeter had to be exploited to its limits. The results were proof that the efforts to obtain higher speed and sensitivity in calorimetric method had not been made in vain. They are a challenge to go further, to be prepared for immunochemical objectives of the future.

7. Thermodynamics of Polynucleotide Helix-formation: Polyribouridylic and Polyriboadenylic Acids

With model-substances for chemical events in which the genetic code is transmitted in living matter, R. Steiner and C. Kitzinger[15] have been first to study certain thermodynamic aspects by direct calorimetry with the heatburst instrument on polyribouridylic and polyriboadenylic acids.

Figure 22 shows the heat of reaction, $\Delta H = -5000$ cal/mole, at a considerable ionic strength, in 0.3 M KCl. In the first part of this article,

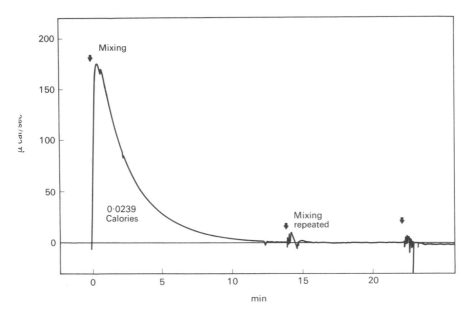

FIG. 22. Heat of formation of polynucleotide helix. Heat of formation of double helices from single-stranded molecules of polyribouridylic and polyriboadenylic acids. The amount of Poly-U was 4.78 μmoles. Poly-A was present in excess. The heat evolved was 23.9 mcal. Note the "zero-spikes" and the reproducibility of the zero baseline in this range.

it has been discussed that the heat of reaction is higher by approximately 800 cal/mole when the ionic strength is low. The single-stranded molecules of Poly-A are then extended, not collapsed. They do not need to be stretched under absorption of heat to form the extended helix. Thus, the observed total heat of reaction is not solely attributable to the formation of two hydrogen bonds per base-pair, between the N_{10} nitrogen of adenine and the C_6 carbonyl of uracil, and between the N_1 of adenine and the N_1 of uracil. One opposite contribution is coil-stretching. Other contributions, imperfectly understood, are assumed to come from Van der Waals forces between adjacent pairs of planar purine and pyrimidine rings.

It should be considered, furthermore, that substantial forces may be required to twist the single strands into the helical conformation. Like in a row of snaps (the hydrogen-bonding sites), one link after another must be formed, in a cooperative process under twisting. One misfitting, non-reactive base-pair might easily interrupt this stepwise procedure.

The forces of two hydrogen bonds in every base-pair may be expected to hold the helices strongly together. Nevertheless, the helices must also come readily apart, and recombine during the genetic process. They do come apart, when rising thermal agitation exerts its disruptive action. This occurs at temperatures near 335°K in observations by Doty.[25]

In reference (26) this phenomenon was utilized to estimate the free energy of helix formation. Since the transition is rather sharp, an equilibrium between double-helices and random coils must exist very near the "melting" temperature, 335°K. At this point, the entropy term, $-T\Delta S$, of the reaction equals the heat change, because the free energy change, ΔF, is zero.

Assuming that the heat, ΔH, at the melting point does not differ substantially from the heat as measured at 298°K, $-T\Delta S$ would equal $\Delta H = -5900$ cal/mole^{-1} at $T = 335$°K. The entropy difference between helices and random coils would be 17.5 units [cal/mole^{-1} deg^{-1}]. At tissue temperature, $T = 310$°K, the entropy term would then be -5420, and the free energy, -480 cal/mole^{-1}.

The low free energy reflects the reversibility of the reaction, a main prerequisite for its function in the genetic process.

In reference (15) the heat of formation of triple helices, Poly $(A + U + U)$ from double helices Poly $(A + U)$, was also measured. This heat, -3500 ± 1000 cal/mole^{-1}, is smaller than the heat of formation of the first double-helix. It has been proposed that the second strand of Poly-U lies in the deep helical groove of the doubly stranded complex and is stabilized by hydrogen bonds between the C_6 carbonyl or uracil and the C_6 amino group of adenine, and between the N_1 nitrogen of uracil and the N_1 nitrogen of adenine. This structure would be formed with a minimal distortion of the bond angles and distances. A later repetition of the heat measurement with triple helix-formation in the heat burst-microcalorimeter by Ross and Scruggs[27] shows excellent agreement with reference (15).

CONCLUSIONS

Principle, instrumentation and procedures of heatburst-microcalorimetry have been described in this chapter to introduce the method to the protein chemist. Concrete examples—a limited selection—were given to illustrate what can be done. The actual scope of the method can only be derived from the universal presence and unlimited reproducibility of heat as a phenomenon of chemical change.

The limitations of the method as well as its wide range are based on its non-specificity. However, specificity of an analytical method is not primarily obtained by specificity of the analytical phenomenon observed. (Optical absorption is only partly specific, and even less specific are weight, volume, pressure, electric current, and other useful signs of reaction.) Specificity of chemical analysis is obtained by experimental design and by proper controls excluding different explanations. Examples of design and control in calorimetry have been described in this article and may be briefly recalled. (1) The previously unknown interaction of chloride-ions with glass surfaces was quantitatively clarified by simply changing amounts of reactants. It was doubly confirmed, by use of vessels of different materials and by saturation of the interface with chloride-ions in control experiments. (2) Heats of neutralization occurring simultaneously with heats of reaction were readily eliminated with control experiments: interaction of protons with the medium in the absence of the reaction under study. (3) Unknown numbers of protons arising or disappearing in reactions were recognized and their heat effects eliminated, by use of buffers with different heats of neutralization. (4) Heats of dilution and ionization were separated from enzymic reaction heat, by graphic analysis of the mixed kinetics. They were eliminated quantitatively by control experiments with reaction products, as shown with ΔH-determination of asparagine hydrolysis. (5) Heat of reaction from a contaminant enzyme was readily distinguished from heat of the reaction under study, by control experiments with products (enzyme-purification). (6) Minute heats of a reverse-reaction (glutamine synthesis) were safely separated from possible side-effects, and properly identified through the identity of the kinetics of the forward and reverse reactions. (7) The heat of stretching of polynucleotide-molecules was observed and separated from the heat of helix-formation by control experiments with varied ionic strength. (8) In measurements of the heat of antigen–antibody interactions, all other sources of heat were ruled out by dual, independent controls – (a) use of serum from a different species and (b) test with a supernatant of the antibody preparation after precipitation of the insoluble antigen–antibody compound. (9) The heat of rearrangement of unevenly distributed surface layers of electrically charged particles on non-conductive solid–liquid interfaces was found and eliminated by observations in metal vessels. (10) Heat, or a deficiency of heat, arising from thermodynamic reversibility, was not only identified with ease. It was utilized for thermodynamic analysis in the glutathione–glyceraldehyde experiment.

There may be numerous other approaches as yet untried or unrecognized, to obtain specific answers in calorimetry. Experimental design and controls defeat the potential disadvantages of non-specificity, while the analytical advantages of heat as a universal phenomenon of chemical change are fully exploited.

ACKNOWLEDGEMENTS
Gratitude is expressed to Dr. Karl Sollner, who influenced this chapter with continuous, constructive criticism. The author is equally indebted to Professor Henry A. Skinner for his most generous advice and criticism,

and to Professors Peter J. W. Debye and Marvin Johnson, and Drs. Koloman Laki and Robert F. Steiner for their most valuable comments.

The work reported here was supported in part under Research Contract No. R-8 and No. R-38 by the National Aeronautics and Space Administration.

REFERENCES

1. W. SWIETOSLOWSKY, *Microcalorimetry*, Reinhold Publishing Corporation, New York, 1946.
2. J. M. STURTEVANT, Heats of biochemical reactions, In: H. A. SKINNER, *Experimental Thermochemistry*, Interscience, New York, 1962.
3. A. V. HILL *et al.*, The heat production of nerves. *Proc. Royal Soc. London* **100**, 223, Ser. B (1926).
4. HAMILTON W. WILSON, *Proc. Phys. Soc., London* **32**,326 (1920).
5. E. CALVET, *Recent Progress in Microcalorimetry*, Pergamon Press, The MacMillan Company, New York, 1963.
6. T. H. BENZINGER and C. KITZINGER, Microcalorimetry of simple biochemical systems. *Fed. Proc.* **13**,11 (A) (1954).
7. T. H. BENZINGER and C. KITZINGER, Microcalorimetry, new methods and objectives. *Temperature–Its Measurement and Control in Science and Industry*, **3**, 3, chap. 5 (1963).
8. T. H. BENZINGER and C. KITZINGER, Direct calorimetry by means of the gradient principle. *The Review of Scientific Instruments* **20**, 849 (1949).
9. T. H. BENZINGER and C. KITZINGER, A four pi radiometer. *Rev. Sci. Instr.* **21**, 599 (1950).
10. O. WARBURG, *Stoffwechsel der Tumoren*, Berlin, 1926.
11. H. HARNED and B. OWEN, *The Physical Chemistry of Electrolytic Solutions*, Reinhold, New York, 1943.
12. C. KITZINGER and T. H. BENZINGER, Waermetoenung der adenosintriphosphorsäure-spaltung. *Z. Naturforsch* **10B**, 375 (1955).
13. K. LAKI and C. KITZINGER, Heat changes during the clotting of fibrinogen. *Nature* **178**, 985 (1956).
14. R. F. STEINER and C. KITZINGER, A calorimetric determination of the heat of an antigen-antibody reaction. *J. Biol. Chem.* **222**, 271 (1956).
15. R. F. STEINER and C. KITZINGER, Heat of reaction of polyriboadenylic acid and poly-ribouridylic acid. *Nature* **194**,1172 (1962).
16. T. B. WEBER, Detection of Extraterrestrial Microorganisms by Microcalorimetry, Final Report (NASA Contract No. NAS-2–2554).
17. T. H. BENZINGER, Equations to obtain for equilibrium reactions, free-energy, heat, and entropy changes from two calorimetric measurements. *Proceedings of the National Academy of Sciences* **42**,109 (1956).
18. C. KITZINGER and R. HEMS, Personal communication (1956).
19. H. A. KREBS, The equilibrium constants of the fumarase and aconitase systems. *Biochemical Journal* **54**,78 (1953).
20. T. H. BENZINGER and R. HEMS, Reversibility and equilibrium of the glutaminase reaction observed calorimetrically to find the free energy of adenosine triphosphate hydrolysis. *Proceedings of the National Academy of Sciences* **42**, 896 (1956).
21. T. H. BENZINGER, C. KITZINGER, R. HEMS and K. BURTON, Free energy changes of the glutaminase reaction and the hydrolysis of the terminal pyrophosphate bond of adenosine triphosphate. *Biochemical Journal* **71**, 400 (1959).
22. L. LEVINTOW and A. MEISTER, *Journal of Biological Chemistry* **209**, 265 (1954).
23. L. KIESOW, On the assimilation of energy from inorganic sources in autotrophic forms of life. *Proceedings of the National Academy of Sciences* **52**, 980 (1964).

24. S. J. SINGER and DAN H. CAMPBELL, Physical chemical studies of soluble antigen–antibody complexes. III. Thermodynamics of the reaction between bovine serum albumin and its rabbit antibodies. *Journal of American Chemical Society* **77**, 3499 (1955).

25. P. DOTY *et al.*, Secondary structure in ribonucleic acids. *Proceedings of the National Academy of Sciences* **45**, 482 (1959).

26. C. KITZINGER, R. F. STEINER and T. H. BENZINGER, Enthalpy changes during the interaction of poly-uridylic and poly-adenylic acids. *Proceedings of the International Union of Physiological Sciences* **2**, 547 (1962).

27. P. D. ROSS and R. L. SCRUGGS, Heat of the reaction forming the three-stranded poly (A and 2U) complex. *Biopolymers* **3**, 491 (1965).

4

PROTEINS AT SEDIMENTATION EQUILIBRIUM IN DENSITY GRADIENTS

By James B. Ifft

from

*Department of Chemistry, University of Redlands,
Redlands, California*

CONTENTS

4

PROTEINS AT SEDIMENTATION EQUILIBRIUM IN DENSITY GRADIENTS

By James B. Ifft

from

Department of Chemistry, University of Redlands, Redlands, California

I. INTRODUCTION

Forty years have passed since Svedberg and Fåhraeus (1926) developed a new technique for the determination of the molecular weights of proteins. The new method consisted of exposing an aqueous protein solution to a very large centrifugal field and allowing the system to come to equilibrium. The equilibrium distribution could then be analyzed optically and the molecular weight ascertained.

The intervening years have been filled with numerous and exciting advances in instrumentation, theory and techniques. Svedberg and Pedersen (1940), in their classic work *The Ultracentrifuge*, detail the advances in theory and instrumentation up to that time. These instruments were custom-made in each laboratory and represented an enormous effort on the part of each investigator just to provide the machinery to begin his experiments. The development of the Model E analytical ultracentrifuge by Beckman Instruments, Inc., Spinco Divison, in 1947 marked the beginning of widespread usage of this technique by protein chemists, until now virtually all laboratories interested in the physical-chemical properties of proteins have at least one Model E in continual use.

Early work in ultracentrifugation centered around two principal techniques, sedimentation velocity and sedimentation equilibrium. As implied in the name, the velocity method involves the measurement of the velocity with which particles move under the influence of the centrifugal field. This velocity is normalized to give the sedimentation coefficient, $S_{20,w}$, which is a characteristic of the polymer at that concentration and in that particular solvent. It is a measure of the molecular weight and the frictional coefficient of the protein. The velocity method also has been extremely valuable as a technique to determine homogeneity.

Sedimentation equilibrium has been used less frequently in the study of proteins because of the relatively long times required to achieve true equilibrium in a normal length column. Archibald's unique insight, in 1947, into the implications of the fact that material never moves through the meniscus or cell bottom, and the demonstration by Van Holde and Baldwin, in 1958, of the value of very short liquid columns have greatly decreased the time needed to obtain useful results by equilibrium measurements.

Claesson and Claesson (1961), in volume 3 of this series, have reviewed carefully the molecular parameters which are available from these classical techniques.

As ultracentrifugation evolved, more complicated problems were explored which required the development of new experimental methods. One of the most important of these depends upon a gradient of density through the liquid column. Such gradients have been utilized in three basically different ways.

One method may be termed stabilized moving-boundary centrifugation (Beckman Instruments, Inc., 1960). The macromolecules are uniformly dispersed throughout the solvent. However, the solvent consists of two components, a highly soluble substance and water. The solvent is inserted into the rotor tube either in layers of decreasing density or as a continuous flow of solvent of decreasing density. In this way, a preformed gradient is established. As the centrifugal field is applied to this solution, the macromolecules migrate outward in the tube through a shallow gradient of density which serves to prevent convection from disturbing the shape of the sedimenting boundary.

A second method, zone centrifugation, overcomes a disadvantage of the moving boundary method, the fact that the faster components of a mixture must sediment through a solution of the slower particles. For zone centrifugation, a preformed solvent gradient is established in the tube. A layer of solution containing the polymer is placed on top of this gradient. The several components of a mixture sediment as discrete zones in this system and full resolution can be achieved. The problem of density destabilization is now even more severe, so that rather steep gradients are required.

The goals of the above two methods are the separation of biological polymers of cellular particles and the measurement of sedimentation constants. A third method, isopycnic gradient centrifugation, has been used historically only for separations. In this method, the gradient of density is selected to encompass the density of the macromolecule (or macromolecules). Particles near the meniscus are in a solvent of lower density than their own and sediment outwards. Particles at the bottom of the tube are less dense than the solvent and float towards the center of rotation. The result is an equilibrium band for each species at a radial position corresponding to its own density. This density will be termed the *buoyant density* in this paper.

Because of the requirement of a preformed gradient, all of the above experiments are performed in preparative ultracentrifuges. This means that an optical record cannot be obtained of the sedimentation process or the equilibrium distribution. Analysis required deceleration of the rotor to rest and a somewhat laborious and difficult examination of the tube contents without permitting redistribution. The gradient in density is generally not known and thus actual values of the buoyant densities of particles have seldom been obtained.

An exciting new variation of the isopycnic technique was described by Meselson, Stahl and Vinograd in 1957. A three-component solution is employed. A small amount of macromolecule is dissolved in a concentrated salt solution of the appropriate density. These experiments differ from the three described above in that this solution is subjected to a centrifugal field which is used to establish the density gradient. Thus, if the analytical ultracentrifuge is employed, the equilibrium concentration distribution can be accurately recorded on film as the solution is rotating. Also because true equilibrium is achieved, a thermodynamic treatment can be employed to provide an exact description of the equilibrium salt distribution. Thus both the shape and position of the band are subject to complete analysis. This method will be termed *sedimentation equilibrium in a density gradient*.

Figure 1 displays the principles of this method. Separations may be achieved, buoyant densities can be accurately determined and, as is

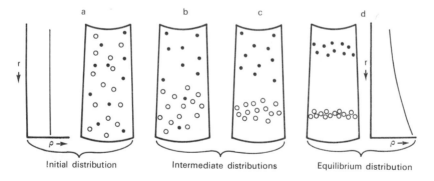

FIG. 1. Schematic diagram of the principle of sedimentation equilibrium in a density gradient. (a) Initially the particles are distributed uniformly in a solution of uniform density. (b, c) As the gradient forms, the particles seek their respective buoyant densities. The larger particles move more rapidly toward their equilibrium position. (d) At equilibrium, complete resolution has been achieved. The band width is inversely proportional to the mass of the particle.

indicated schematically, the breadth of the distribution can yield the molecular weight of the species.

Since the development of this technique in 1957, there has appeared an enormous body of literature dealing with the results of experiments with RNA and DNA in preparative and analytical ultracentrifuges. The preparative experiments have been designed primarily to separate the nucleic acids and they frequently have not been true equilibrium runs. The analytical runs have been at equilibrium and yield values of the buoyant density. They reveal minute differences in the buoyant densities of nucleic acids derived from different parts of the cell or due to isotopic labelling. (See Mahler and Cordes (1966) for a recent, brief discussion of some of the applications of density gradient centrifugation to nucleic acids.)

A corresponding effort has not been expended on the examination of proteins at sedimentation equilibrium in density gradients. A sizeable portion of work has been done to utilize the preparative centrifuge as a tool to isolate proteins in fairly pure form. An example of this is the work of Anderson and his group at Oak Ridge. They have prepared an extremely thorough monograph (National Cancer Institute Monograph, No. 21, 1966) which describes the development of continuous flow centrifuges as tools to prepare large quantities of cellular particles, viruses and biological macromolecules in relatively pure form for further biological studies. Also an extensive account of preparative centrifugation involving density gradients has been given by deDuve *et al.* (1959). An excellent brief summary and extensive bibliography of this same topic is provided in Spinco's little Technical Bulletin No. 1 (Beckman Instruments Co., Inc., 1960). Trautman (1964) has given an excellent review of the entire subject of ultracentrifugation. Included in his chapter are numerous informative figures as well as descriptions of preparative procedures. Because of these recent reviews, preparative centrifugation involving density gradients will be discussed only briefly in the present paper.

This paper will deal primarily with the application of the technique of sedimentation equilibrium in density gradients as an analytical method which seeks to determine physical-chemical properties of proteins. Properties which are presently accessible with this method are the molecular weight, buoyant density, solvation and ion binding of a protein. The theoretical work which has been done to describe this system almost exceeds the experimental work. The requisite theoretical background is presented in Section II. A detailed description of how to conduct these experiments is given in Section III. A brief description of preparative techniques is given in this section and some preparative results are presented along with the analytical applications which are given in Section IV.

II. THEORY

A.   Distribution of the Two-component Solvent

1. Equation for Two-component Sedimentation Equilibrium

The distribution of a solute through the centrifuge cell at sedimentation equilibrium can be obtained from either a thermodynamic or a kinetic point of view. Because the former method requires almost no assumptions and the nature of the variables such as density is not open to question, we will consider this derivation.

The thermodynamic requirement for a system to be at equilibrium at every point in the phase is that the total potential, ξ, be the same throughout the phase. The dependence of this potential on the several thermodynamic parameters is:

$$\xi = \xi\,(T, P, a, r), \tag{1}$$

where T is the absolute temperature, P is the pressure, a is the activity of the solute and r is the radial distance from the center of rotation.

The total derivative of this function with respect to distance is given by:

$$\frac{d\xi}{dr} = \left(\frac{\partial\xi}{\partial T}\right)_{P,a,r}\frac{dT}{dr} + \left(\frac{\partial\xi}{\partial P}\right)_{T,a,r}\frac{dP}{dr} + \left(\frac{\partial\xi}{\partial a}\right)_{T,P,r}\frac{da}{dr} + \left(\frac{\partial\xi}{\partial r}\right)_{T,P,a}. \tag{2}$$

This derivative is identically zero at every point in the cell when the system is at equilibrium. Because virtually all centrifuge experiments are conducted isothermally, the first term is equal to zero.

All of the other derivatives are well known and are listed here:

$$\left(\frac{\partial\xi}{\partial P}\right)_{T,a,r} = \bar{V},$$

$$\frac{dP}{dr} = \rho\omega^2 r,$$

$$\left(\frac{\partial\xi}{\partial a}\right)_{T,P,r} = \frac{RT}{a},$$

and

$$\left(\frac{\partial\xi}{\partial r}\right)_{T,P,a} = -M\omega^2 r,$$

where $\bar{V}$ is the partial molar volume of the solute, ρ is the density of the solution, ω is the angular velocity in radians/sec and M is the molecular weight of the solute. Noting that $\bar{V} = M\bar{v}$, where $\bar{v}$ is the partial specific volume of the solute, Eq. (2) reduces to:

$$\frac{da}{dr} = \frac{M(1-\bar{v}\rho)\omega^2 r}{RT} \cdot a. \tag{3}$$

This is the classic equation of two-component sedimentation equilibrium. It has been derived in a more general and elegant fashion in a classic paper by Goldberg (1953).

2. Calculation of the Density and Density Gradient Distributions

The distribution of salt within the centrifuge cell at equilibrium has been computed in several ways. Ifft, Voet and Vinograd (1961) employed the relation:

$$\frac{d\rho}{dr} = \frac{d\rho}{da} \cdot \frac{da}{dr} = \frac{d\rho}{da} \cdot \frac{M(1-v\rho)\omega^2 r}{RT} \cdot a = \frac{d\rho}{d\ln a} \cdot \frac{M(1-\bar{v}\rho)\omega^2 r}{RT}.$$

We define β^0 such that:*

$$\frac{d\rho}{dr} = \frac{\omega^2 r}{\beta^0}. \tag{4}$$

This density gradient will be referred to as the composition density gradient because it arises solely from the redistribution of the solute. The quantity β^0 was tabulated as a function of density by graphical methods from density and activity data. The results of these calculations are given in Table I. Thus, if the density is known at any position r in the cell, the composition density gradient can be computed for any angular velocity.

The density distribution can now be directly computed from the differential equation:

$$\beta^0(\rho)d\rho = \omega^2 r dr. \tag{5}$$

Third-order polynomials were obtained for $\beta^0(\rho)$. The coefficients of these polynomials for five solutes are given in Table II.

Equation (5) was integrated and solved by a digital computer for density as a function of distance utilizing the appropriate conservation of mass relation for sectorial or cylindrical cells. Radial distances for

*A superscript zero will be used in this paper to denote variables defined at atmospheric pressure. Thus the symbol, β^0, will be employed throughout this paper to emphasize that the values of β obtained by Ifft, Voet and Vinograd are strictly valid only at 1 atm.

TABLE I. VARIATION OF $\beta^0 \times 10^{-9}$ WITH SOLUTION DENSITY

ρ	Sucrose	KBr	RbBr	RbCl	CsCl
1.02	8.091				
1.03	6.789				
1.04	5.605				
1.05	4.643		6.729	9.817	
1.06	4.019				
1.075		7.496			
1.08	3.449				
1.10	3.237	6.121	3.643	5.532	
1.12	3.121				
1.125		5.229			
1.14	3.091				
1.15		4.594	2.536	4.109	2.491
1.175		4.151			
1.20		3.848	2.122	3.445	1.984
1.225		3.637			
1.250		3.469	1.772	3.172	1.715
1.275		3.330			
1.30		3.213	1.635	3.083	1.546
1.325		3.112			
1.35			1.528	2.777	1.430
1.40			1.434	2.334	1.346
1.45			1.372		1.286
1.50					1.245
1.55					1.216
1.60					1.197
1.65					1.190
1.70					1.190
1.75					1.199
1.80					1.215
1.85					1.236

the meniscus and cell bottom were chosen to correspond to a 0.70-ml filling of a 12-mm, 4° sector or a 2-, 3- or 5-ml filling in 5-ml plastic tubes employed in the SW-39 swinging-bucket rotor which is used in the Spinco Model L preparative centrifuge. The results obtained were the densities at either 5 or 7 evenly spaced positions in the cell or tube and the iso-concentration* position, r_e, that is the radial position at which the

---

*The term isoconcentration is preferred rather than isodensity to emphasize that this quantity depends only upon the composition of the solution and is independent of pressure effects.

TABLE II. POLYNOMIAL COEFFICIENTS IN THIRD-ORDER POLYNOMIALS
USED TO DESCRIBE $\beta^0(\rho)$

$$\beta^0(\rho) = \beta_0 + \beta_1\rho + \beta_2\rho^2 + \beta_3\rho^3.$$

Salt	Density range	$\beta_0 \times 10^{-9}$	$\beta_1 \times 10^{-9}$	$\beta_2 \times 10^{-9}$	$\beta_3 \times 10^{-9}$
CsCl	1.15–1.85	47.9726	−85.3724	51.7764	−10.4312
RbBr	1.05–1.45	620.9619	−1416.8024	1079.2080	−273.7280
RbCl	1.05–1.40	1222.1450	−2878.2225	2264.0729	−593·3408
KBr	1.075–1.325	966.7593	−2280.4334	1802.8082	−475.9704
Sucrose	1.02–1.14	860.6168	−2320.1077	2085.6993	−625.0000

density of the solution is the same as the initial, uniform solution density, ρ_e. Some representative density distributions are given in Figs. 2, 3 and 4. As expected, the distributions are steeper for solutions having higher initial ρ_e's and for higher ω's. The maximum density difference which can be obtained with CsCl in a 5-ml cylinder in an SW-39 rotor at 39,000 rpm is 0.4 g/ml.

To further systematize the results and simplify the use of this data, the concept of normalized isoconcentration points, $(r_e - r_a)/(r'_e - r_a)$ was employed where r_a is the radial position of the meniscus. It was recognized that if $\beta(\rho)$ is a constant, limiting isoconcentration positions,

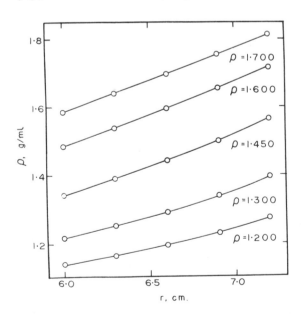

FIG. 2. Density distributions for CsCl solutions at 25.0°C at equilibrium in an analytical ultracentrifuge cell at 56,100 rpm. (Ifft *et al.*, 1961.)

r_e' can be derived as:

$$r_e' = \sqrt{\left(\frac{r_a^2 + r_b^2}{2}\right)} \quad \text{(sectors)},$$

$$r_e' = \sqrt{\left(\frac{r_a^2 + r_a r_b + r_b^2}{3}\right)} \quad \text{(cylinders)},$$

where r_b is the radial position of the bottom of the liquid column. The deviation of the normalized isoconcentration values from 1.000 can be used as a criterion to determine whether the easily computed values of r_e'

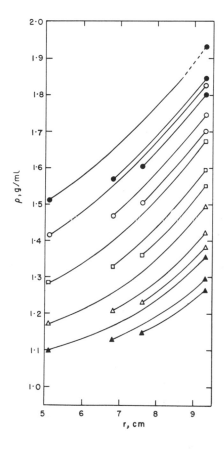

FIG. 3. Density distributions for CsCl solutions at 25.0°C at equilibrium in cylindrical tubes containing 2, 3 and 5 ml of solution in a preparative ultracentrifuge at 39,000 rpm: ●, $\rho_e = 1.7$; ○, $\rho_e = 1.6$; □, $\rho_e = 1.45$; △, $\rho_e = 1.3$; ▶, $\rho_e = 1.2$. (Ifft $et\ al.$, 1961.)

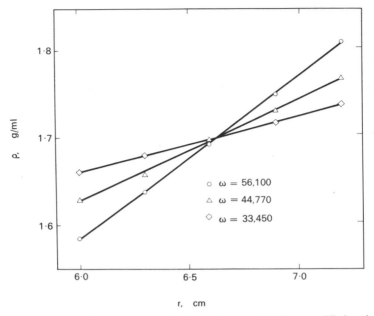

FIG. 4. Density distributions for CsCl solutions at 25.0°C at equilibrium in an analytical ultracentrifuge cell as a function of angular velocity, $\rho_e = 1.7$. (Ifft *et al.*, 1961.)

can be used instead of the more accurate tabulated values of $(r_e - r_a)/(r'_e - r_a)$. It is clear from Fig. 5, which presents some of these data, that for proteins in the density range of 1.3, errors between 2 and 6% will result in r_e if values of r'_e are used in its place.

The authors concluded with a discussion of considerable practical significance. They noted that a straight line through successive points on the $\beta(\rho)$ curve represents the actual curve better than the polynomial curve does. They showed that the following relation is good for sectors to better than ±1%:

$$(r'_e)^2 - (r_e)^2 = D(r_b^2 - r_a^2)^2/48, \tag{6}$$

where $D = g\omega^2/\beta^2$ and $g =$ slope of the $\beta(\rho)$ curve at ρ_e. Values of r_e obtained by this method were in general within 0.005 cm of the computer values. This is an acceptable agreement. The importance of Eq. (6) therefore is that accurate isoconcentration positions in sectors can be computed without recourse to a computer by making large-scale plots of the β data in Table I and graphically determining slopes at the densities of interest. The entire density distribution then can be computed by a stepwise integration using the isoconcentration position and the $\beta(\rho)$ data.

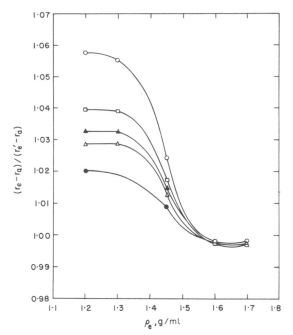

FIG. 5. Normalized isoconcentration points as a function of original solution density for CsCl at 25°C: ○, 5-ml cylinder, 39,000 rpm; □, 3-ml cylinder, 39,000 rpm; △, 2-ml cylinder, 39,000 rpm; ▲, sector, 56,100 rpm; ●, sector, 44,770 rpm. (Ifft *et al.*, 1961.)

Trautman (1960) has proposed a more accurate method of computation of the β-function than that given above. His procedure differs from that described in that he deals with activity coefficients rather than activities. The activity coefficient on the conventional molality concentration scale, γ, is converted to the activity coefficient on a density scale, γ'. Then the reciprocal of β is computed as:

$$\alpha = \frac{1}{\beta} = \frac{M(1 - \bar{v}\rho)(\rho - \rho_0)}{\nu RT} \frac{1}{[1 + \partial \ln \gamma'/\partial \ln (\rho - \rho_0)]}$$

where ν is the number of ions per molecule and ρ_0 is the density of the solvent. Trautman's data have been converted to β values and are presented in Fig. 6. These date agree with that of Ifft, Voet and Vinograd within 1% throughout the solubility range of CsCl. They can be directly compared via the linear relation (Hearst and Vinograd, 1961b):

$$\text{Wt.\% CsCl} = 137.48 - 138.11 \left(\frac{1}{\rho^{25}} \right). \tag{7}$$

Trautman notes that his method is much more accurate than that based on activities in the concentration range less than 10% because the slope of

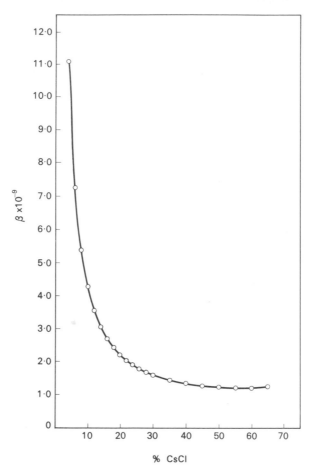

FIG. 6. Density gradient proportionality constant β as a function of density for CsCl, 25°C. Calculated as the reciprocal of the data given by Trautman (1960).

the ln a vs. ρ plot is much greater than that of the ln γ' vs. ln $(\rho - \rho_0)$ plot.

A third method of computation of both the density and density gradient distributions has recently been given by Ludlum and Warner (1965). Their method is based upon the usage of the molal osmotic coefficient, ϕ. The parameter β is evaluated as:

$$\beta = \frac{3RT\,[\,(\phi/m) + (\mathrm{d}\phi/\mathrm{d}m)\,]}{(M - \bar{V}\rho)\mathrm{d}\rho/\mathrm{d}m}\,,$$

where m is the molality of the solution.

The density distribution is evaluated by expressing the density at any point via a Taylor's expansion. This equation and Eq. (4) of this paper are successively differentiated and the coefficients of the Taylor's expansion are thus determined as derivatives of $\ln \beta$ with respect to ρ. The density at any point in the cell is now determined by invoking the conservation of mass relation, guessing approximate values of ρ_i at r_i and arriving at the correct ρ_i by an iterative procedure.

The results for β and the derivatives of $\ln \beta$ with respect to density as functions of density are presented in Fig. 7* for Cs_2SO_4 solutions.

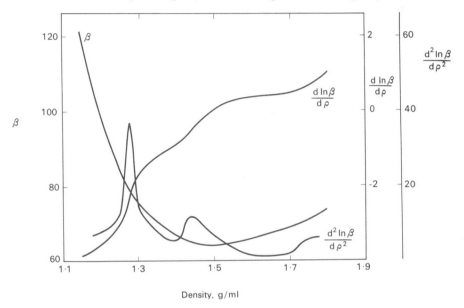

Density, g/ml

FIG. 7. Density gradient proportionality constant β (Cs_2SO_4) and derivatives of $\ln \beta$ with respect to density as functions of density. $\beta \times 10^{-7}$ is given in $cm^5\ g^{-1}\ sec^{-2}$. (Ludlum and Warner, 1965.)

In order to obtain the above relations, the authors had to accurately determine values over the entire solubility range of Cs_2SO_4 for m, the molality, ρ, ϕ and $(n-n_0)$, the difference in refractive index between solution and solvent. Two relations are presented which are of considerable practical use:

$$\rho = 1.0047 + 0.28369m - 0.017428m^2\ (0.5 \leqslant m \leqslant 3.6),$$

$$\rho = 0.9954 + 11.1066(n-n_0) + 26.4460(n-n_0)^2.$$

$$(1.14 < \rho < 1.80).$$

*Dr. Warner has indicated that large-scale copies of Fig. 7 may be obtained from the Department of Biochemistry, New York University School of Medicine, New York, New York 10016.

Approximate values of $1/\beta$ at 20°C for cesium formate, cesium acetate, rubidium formate, potassium acetate, potassium tartrate, sodium formate and lithium chloride have been computed by Hu *et al.* (1962).

3. Effects of Pressure on the Density and Density Gradient Distributions

All of the above treatments apply to binary solvents in a centrifugal field in which the pressure is 1 atm. However, the hydrostatic pressures in the centrifuge cell vary from 1 atm at the meniscus to several hundred atmospheres at the bottom of the cell.

Hearst *et al.* (1961) have considered the effects of pressure on a two component system at sedimentation equilibrium. The equation,

$$M_2(1 - \bar{v}_{2,P}\rho)\omega^2 r\mathrm{d}r = \left(\frac{\partial \mu_2}{\partial m_2}\right)P\mathrm{d}m_2, \tag{8}$$

obtained by Goldberg (1953) is a more general relation than Eq. (3) to describe a two-component system at equilibrium in a centrifugal field. The subscript 2 denotes the solute. The functions $\bar{v}_2$, ρ and $(\partial\mu_2/\partial m_2)$ were expanded as first-order Taylor series in terms of pressure. This directly yields the expression required to convert densities applicable to 1 atm to the solution at a pressure P as:

$$\rho_i = \rho_i^0(1 + \kappa P_i), \tag{9}$$

where P_i is the pressure at point i and κ is the isothermal compressibility coefficient.

Substitution of these three expansions into Eq. (8) and evaluation of the magnitudes of the corrections involved reveals that the composition density gradient is still correctly formulated by Eq. (4). This result, that pressure does not affect the equilibrium distribution of salt, is both surprising and pleasing. It means that the density, ρ^0, which depends only upon the composition of the solution, m_2, still can be computed correctly from the previously tabulated values of β^0 as:

$$\rho^0 = \rho_e^0 + \int_{re}^{r} \left(\frac{\omega^2 r}{\beta^0}\right)\mathrm{d}r. \tag{10}$$

The physical density gradient is defined to be the actual gradient in density which exists at any point in the cell and is given the symbol, $(\mathrm{d}\rho/\mathrm{d}r)_{phys}$. It is obtained by differentiation of Eq. (9) and is expressed as:

$$\left(\frac{\mathrm{d}\rho}{\mathrm{d}r}\right)_{phys} = \left[\left(\frac{1}{\beta^0}\right) + \kappa\rho^{0^2}\right]\omega^2 r. \tag{11}$$

For the case in which there is no preferential interaction of the solute with the macromolecule, this is the proper density gradient to use in the calculation of the molecular weight from Eq. (13).

B.   Ideal Behavior of Macromolecules in a Density Gradient

1. Qualitative Description of Factors Determining Buoyant Density

The most readily and most frequently measured quantity in density gradient experiments is the buoyant density, ρ_0^0, which is the density of the solution at band center resulting solely from the composition of the solution. Before we approach the problem in a quantitative, thermodynamic manner, it may prove useful to consider qualitatively the physical factors which determine the buoyant density.

The most important factor to consider is the composition of the macromolecule. Consideration of the densities of some molecules of interest reveals some interesting differences (see Table III). The densities of the

TABLE III. DENSITIES OF MOLECULES
OF INTEREST

Compound	Density
Glycerol	1.261
Oleic acid	0.895
Glycerol trioleate	0.915
Glycerol mononitrate	1.40
Glucose	1.562
Glucose, 2-methyl-(D)	1.46
L-Aspartic acid	1.66
L-Alanine	1.401
L-Arginine	1.1
Most dry proteins*	1.27
CsCl	3.988
H_2O	0.997

* Haurowitz (1963a).

monomers indicate that we could arrange the polymers in order of increasing density as lipids, proteins and polysaccharides. Although the densities of nucleotides are not readily available, we know experimentally that the nucleic acids have higher densities than the preceding three classes of biological compounds.

Haurowitz (1963a) has given the density of most dry proteins as 1.27 g/ml. However, as is indicated by the wide range of densities exhibited by the amino acids in Table III, densities ranging from 1.2 to 1.5 might be expected from a simple weighting according to amino acid composition. As would be expected, the density of a molecule formed by condensation of two molecules is intermediate in density between the two reacting molecules. This effect is observed in Table III for the addition of a low-density hydrocarbon to glucose or glycerol. Thus, we would expect the densities of conjugated proteins such as lipoproteins to be less than that of a pure protein and the densities of glucoproteins and nucleoproteins to be above this average value for pure proteins due to the differing densities of the prosthetic groups.

These observations as to predicted densities of proteins in solution are gross approximations for a number of reasons. Proteins are formed by dehydration reactions between amino acids. Examination of amino acid densities alone neglects the less-dense water molecule which is lost and does not take into account the contribution of the peptide bond to the density. In addition, proteins are known to interact extensively with small molecules in solution (Haurowitz, 1963b). As is apparent from the effect of nitration on the density of glycerol and from the listed densities of water and a typical salt employed in these experiments, hydration and ion-binding will respectively lower or raise the observed buoyant density. In view of these significant second-order interactions, it is somewhat surprising that the buoyant densities of all non-conjugated proteins reported to date fall in the range 1.20–1.35.

2. Fundamental Relation Describing the Macromolecular Band

Meselson *et al.* (1957) developed the fundamental equation for non-interacting systems. They began with the fundamental relation for sedimentation equilibrium, Eq. (8). Substitution of a first-order Taylor expansion involving the density gradient, $d\rho/dr$, for the density and the good assumption that the distance from the center of the band to any point in the band is small compared to the distance from the center of rotation leads directly to the important relation:

$$c = c_0 e^{-(r-r_0)^2/2\sigma^2}, \tag{12}$$

where

$$\sigma^2 = \frac{RT}{M\bar{v}(d\rho/dr)_{r_0}\omega^2 r_0}. \tag{13}$$

The quantities c, $\bar{v}$ and M refer to the concentration, partial specific volume and molecular weight of the unbound macromolecule. The result is extremely interesting because it demonstrates that for an ideal system, the distribution of macromolecules in the band is Gaussian of standard deviation, σ. Meselson *et al.* (1957) experimentally demonstrated the validity of Eq. (12) by recording the absorption of light through a band of T-4 DNA and comparing it with a computed Gaussian curve. As seen in Fig. 8, the agreement is excellent.

The authors present equations which demonstrate how either weight-average, M_w, or number-average, M_n, molecular weights can be obtained from analyses of equilibrium bands. The calculations involve integrations through the entire band.

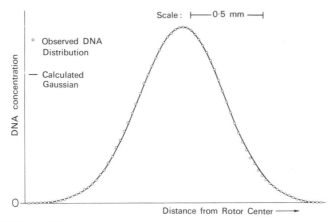

FIG. 8. The equilibrium distribution of DNA from bacteriophage T-4 at 27,490 rpm in 7.7 molal CsCl at pH 8.4. $\rho_0^0 = 1.70$ g/ml; $(d\rho/dr)_{comp} = 0.046$ g/cm^4. (Meselson *et al.*, 1957.)

3. Time Required to Reach Equilibrium

The problem of computing when equilibrium is achieved in a density gradient experiment has been considered by Meselson and Nazarian (1963). Their treatment assumes that the gradient has been established and is constant as the macromolecules seek their equilibrium positions. The time required to approach to within 1% of the equilibrium distribution is:

$$t^* = \frac{\sigma^2}{D}\left(\ln\frac{L}{\sigma}+1.26\right), \tag{14}$$

where D is the diffusion coefficient of the polymer and provided that the length of the liquid column, L, is very much larger than the standard deviation of the Gaussian band, σ. It is seen that the time required to approach equilibrium within 1% depends only weakly on the length of the liquid column because the macromolecule is contained in a narrow band but t^*, according to Eqs. (14) and (31), depends inversely on the fourth power of the angular velocity. The value of t^* for serum albumin with $D = 5.94 \times 10^{-7}$ (Tanford, 1961), $\sigma = 0.10$ cm and a full analytical cell of column length 1.2 cm is 17.5 hr.

The time required to achieve sensible equilibrium in a two-component system has been examined by Van Holde and Baldwin (1958). We may use their results to compute the time required for the equilibrium salt distribution to be achieved. The fundamental relation which they provided is:

$$t_\epsilon = (r_b - r_a)^2 F(\alpha, \epsilon)/D.$$

$F(\alpha, \epsilon)$ is a function of the parameters $\alpha = 2RT/M(1 - \bar{v}\rho)\omega^2(r_b^2 - r_a^2)$ and ϵ which is a measure of the departure from equilibrium. The diffusion coefficient of the solute is given by D and r_a and r_b have been defined above (Eq. (6)). It is apparent from this relation that the time required for the solute to reach its equilibrium distribution depends strongly on the column length and is almost independent of the centrifugal field.

The parameter α for a CsCl solution of density 1.30 at 56,100 rpm in a full cell at 25°C is 0.82. This corresponds to a value of $F(\alpha)$ of 0.66 if $\epsilon = 0.001$. Lyons and Riley (1954) have given the diffusion coefficients of CsCl solutions over the entire concentration range. Insertion of their value of 2.19×10^{-5} for D at density 1.30 yields a value of $t_{0.001} = 133$ hr. It will be apparent from the times quoted in the experimental section that most density gradient experiments need not proceed to within 0.1% of the equilibrium distribution before no further detectable changes are noted in the salt or in the macromolecular distribution.

4. Resolution

Several authors have considered the problem of the power of the density gradient method to physically separate materials of differing density. Ifft et al. (1961) defined resolution, Λ, as:

$$\Lambda = \frac{\Delta r}{(\sigma_1 + \sigma_2)}. \tag{15}$$

Lambda should be a large number for materials which are separated by a large distance, Δr, and which have small standard deviations. Combination of Eqs. (4), (13) and (15) and suitable approximations yields:

$$\Lambda = \Delta\rho\sqrt{1/RT}\{\sqrt{(M_1M_2)}/(\sqrt{M_1} + \sqrt{M_2})\}\sqrt{\{(\beta/\rho)_{r_0}\}}, \tag{16}$$

where $\Delta\rho$ is the difference in the two buoyant densities and β and ρ are evaluated at the mean of the two banding positions, $\bar{r}_0$. This equation is further simplified to:

$$\Lambda = \frac{\Delta\rho}{2}\sqrt{\{(M\beta'/\rho)_{\bar{r}_0}\}}, \tag{17}$$

for the case that the molecular weights are about the same and $\beta' = \beta/RT$.

Equation (17) provides the interesting result that resolution is independent of angular velocity. The bands become narrower as ω is increased but the separation Δr decreases. The equation further demonstrates that resolution can be improved by selection of a salt with a large value of β. The data of Table I indicate that KBr and RbCl will resolve about 1.4 times better than CsCl. There are two ways to interpret the meaning of a given value of Λ. If $\Lambda = 1$, overlap of the two bands will be so large that only one maximum in the concentration distribution will be observed. At $\Lambda = 2$ there is about 5% intermixing, and at $\Lambda = 3$ the materials are virtually resolved.

An analytical method can also be derived for examining this separation. If we define y to be the ratio of the sum of the concentrations of both species midway between band centers to the concentration at the maximum of each distribution, this elevation is given by the relation:

$$y = 2 \cdot \exp(-\Lambda^2/2). \tag{18}$$

Equation (18) is applicable only if the initial concentrations and standard deviations are equal. Figure 9 gives a plot of y vs. Λ to show how sensitive the separation is to the resolution parameter.

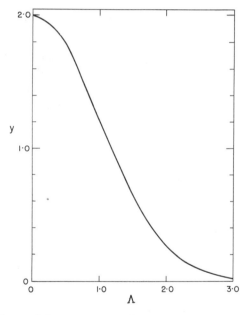

FIG. 9. Dependence of the parameter, y, the ratio of the concentrations midway between two modes to the concentration at the mode on the resolution parameter Λ for two materials of nearly the same molecular weight present at equal concentrations (Vinograd and Hearst, 1962.)

Hu *et al.* (1962) defined resolution as above and calculated several plots which are of considerable practical interest. The first plot, Fig. 10, provides the resolution attainable in RbCl solutions for a given density

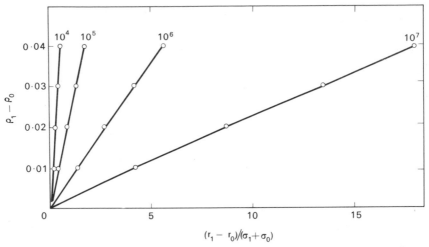

FIG. 10. Resolution $\Lambda = (r_1 - r_0)/(\sigma_1 + \sigma_0)$ as a function of the difference in buoyant densities of two species. Data apply to RbCl solutions of $\rho_e = 1.33 \text{ g/ml}$ for proteins having the molecular weights indicated on the curves. (Hu *et al.*, 1962.)

difference of macromolecules of molecular weights given on the lines. The second plot, Fig. 11, gives the expected distance separation between modes as a function of angular velocity for several salts. While this plot does not provide any information regarding resolution, it is useful in knowing where to look for bands when a preparative gradient column is being analyzed.

C. Real Behavior of Macromolecules in a Density Gradient

The meaning of the buoyant density and the method of calculation of and the interpretation of a molecular weight has been under vigorous investigation since the pioneering work of Meselson, Stahl and Vinograd. Much of this discussion has been oriented toward an understanding of the behavior of nucleic acids in a salt gradient. Nevertheless, we can apply significant portions of these theoretical advances to an understanding of the behavior of proteins.

1. Effect of Density Heterogeneity

Meselson *et al.* (1957) examined several cases of density heterogeneity. If the densities are sufficiently far apart, a bimodal or polymodal distribution will result. A skewed band would result from molecules exhibiting slightly differing densities. A band which is symmetrical but whose $\ln c$ vs. $(r - r_0)^2$ plot is not linear consists of material which is either heterogeneous with respect to density (concave upward) or molecular weight (concave downward). It is highly unlikely that a Gaussian band could

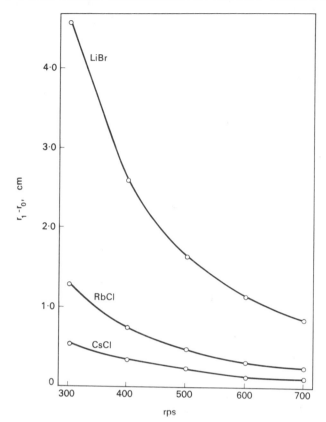

FIG. 11. Equilibrium band separations as a function of angular velocity for several salt solutions. Data apply to use of a Spinco SW-39 rotor, rps refers to revolutions per second and the buoyant densities are 1.33 and 1.34 g/ml. (Hu *et al.*, 1962.)

result from a combination of density and molecular weight heterogeneity.

Baldwin (1959) considered the special case of a Gaussian distribution of densities. A Gaussian concentration distribution will still result but a standard deviation of densities as small as 0.003 g/ml will yield an error of 2 in the determination of the molecular weight.

A more general analysis, applying to any distribution of molecular weights or densities, is given by Sueoka (1959). The total variance of the distribution is expressed as the sum of the variances due to thermal motion and density heterogeneity. A lengthy analysis of the sedimentation boundary is required to utilize this method.

Baldwin and Shooter (1963) have suggested the use of preformed density gradients to separate the contributions of the variances of macromolecular densities and molecular weights to the standard deviation of the band. This approach has an advantage over that of Hearst (1961) in that the same solvent is employed in each experiment. However, there is the severe experimental problem of selecting precisely the correct initial preformed gradient.

2. Effect of Charge

Yeandle (1958) has examined the effect of each nucleotide residue of molecular weight 330 of DNA having unit negative charge which is totally uncompensated. Inclusion of this electrical field in the derivation indicates that an error of 15% will result in the determination of the molecular weight. Because it is likely that primary cation binding will leave only a small residual negative charge, this effect is not likely to be appreciable, even for nucleic acids. It should be even less for proteins, of which only a fraction of the residues can ionize.

3. Effect of Concentration

Dayantis and Benoit (1964a, 1964b) considered the effect of the variation of the activity coefficient of the polymer with its concentration. They determined the relationship between the apparent molecular weight and the actual molecular weight as a function of the method of computation employed. For instance, measurement of the distance between the inflection points to yield σ yields the relation:

$$\frac{1}{M_{app}} = \frac{1}{M} + 2.42\, A_2 c_m,$$

where A_2 is the second virial coefficient and c_m is the concentration at the center of the band. Because the latter two variables are relatively small for typical experiments with proteins and because the precision of measurement is not very high at present, no concentration dependencies have been reported for proteins as these authors report for polystyrene in a mixed solvent of methylethyl ketone and chloroform.

4. Effect of Temperature

The only report on temperature variation studies in density gradients involving macromolecules is that by Vinograd et al. (1965). The buoyant density of DNA was measured from 5 to 65°C. They found a buoyant density increase which was ascribed to a loss of water of hydration. The important point to note from their work is that the thermal expansion coefficient, γ_{CsCl}, for the CsCl concentration employed must be used to compute the correct solution densities. These values are available in the International Critical Tables (1928).

5. Effect of Polydispersity

This variable has already been alluded to in the treatments of Meselson et al. (1957) and Sueoka (1959).

Ende (1965) has done a thorough study of the problem of which ratio of molecular weight averages obtained from density gradient analyses is the least concentration dependent. He concludes that the ratio of M_w, which is obtained from the distance between the inflection points, and a strange average $<M>_{5/2}$, which is obtained from σ^2 and approaches M_z under certain conditions, should be used. Although most protein preparations are far more homogeneous than the polystyrene samples Ende was studying, his results should be considered if there is evidence of any proteolytic cleavage in a particular protein preparation.

6. Effect of Pressure

The effect of pressure on the salt distribution in the cell under operating conditions of several hundred atmospheres has already been given in Eq. (9) and (11). The relation between ρ_0^0 and P will be discussed in the following section.

7. Effect of Preferential Interaction

The binding by the polymer of one or both components of the solute is the most important, the most difficult to treat and the most thoroughly researched of any of these secondary factors we have been considering. Because the salt is much more dense and water is lighter than proteins, we would expect that the observed buoyant density may deviate markedly from the dry density of the protein. Binding of water and/or salt will increase the molecular weight. To further compound the difficulty, it should be apparent that the physical density gradient given in Eq. (11) will not give the correct density gradient to use in Eq. (13) to determine the molecular weight if either the compressibility of the polymer differs from that of the solution or if the solvation of the polymer changes through the band.

Williams *et al.* (1958) were the first to consider the effect of solvation on the buoyant density of the macrospecies. They began with Goldberg's equations for sedimentation equilibrium in a three-component system* at constant temperature and pressure:

$$M_1(1-\overline{v}_1\rho)\omega^2 r\, dr = \left(\frac{\partial\mu_1}{\partial m_1}\right)_{m_3} dm_1 + \left(\frac{\partial\mu_1}{\partial m_3}\right)_{m_1} dm_3, \qquad (19a)$$

$$M_3(1-\overline{v}_3\rho)\omega^2 r\, dr = \left(\frac{\partial\mu_3}{\partial m_1}\right)_{m_3} dm_1 + \left(\frac{\partial\mu_3}{\partial m_3}\right)_{m_1} dm_3. \qquad (19b)$$

---

*In this chapter, we will use the following notation: 1 represents water, 2 the salt and 3 the macromolecule.

These expressions are the rigorously correct analogues of Eq. (3) and (8). Division of (19a) by (19b), noting that at band center $(dm_3/dr) = 0$ and identification of the solvation Γ as $-(\partial\mu_1/\partial m_3)_{m_1}/(\partial\mu_1/\partial m_1)_{m_3}$ yields the following relation between Γ' the solvation in grams water/gram polymer and the buoyant density:

$$\frac{1}{\rho(r_0)} = \frac{\bar{v}_3 + \Gamma'\bar{v}_1}{1 + \Gamma'}. \tag{20}$$

This relation is intuitively reasonable in that it asserts that the density of a solvated particle is equal to the mass of the particle divided by its volume. As is apparent from the derivation, the density to be employed is the density at band center corrected for the effects of pressure.

Baldwin (1959) has extended this treatment by expanding the variables $\rho, \bar{v}_3, \bar{v}_1$ and Γ' in a first-order Taylor series and arrived at the approximate expression:

$$\sigma^2 = RT/(M_3\omega^2 r_0)\left[(\bar{v}_3 + \lambda'\bar{v}_1)\frac{d\rho}{dr} - \left(1 - \bar{v}_1\rho\right)\left(\frac{\partial\lambda'}{\partial c_1}\right)_P\frac{dc_1}{dr}\right]r_0. \tag{21}$$

The left-hand side is the square of the standard deviation of the Gaussian concentration distribution, $\bar{v}_1$ and $\bar{v}_3$ are partial specific volumes of one of the solutes and the macromolecule respectively and λ' again is the solvation. The derivation requires the neglect of two terms, $d\bar{v}_1/dr$ and $d\bar{v}_3/dr$, the second of which is not likely to be negligible.

The effect of solvation has been examined also by Hearst and Vinograd (1961a). They began with Goldberg's relations, Eqs. (19). This led directly without any assumptions to the same expression for solvation which Williams et al. had obtained (Eq. (20)). Invoking a thermodynamic criterion for stability with respect to new phase formation, the following relation was obtained:

$$M_s(1 - \bar{v}_s\rho)\omega^2 rdr = \frac{RT}{m_3} dm_3, \tag{22}$$

where

$$M_s = M_3(1 + \Gamma') \quad \text{and} \quad \bar{v}_s = (\bar{v}_3 + \Gamma'\bar{v}_1)/(1 + \Gamma').$$

The effect of solvation on the density gradient was considered and v_s, ρ and M_s expanded about band center to yield:

$$\sigma^2 = \frac{RT}{M_{s,0}\bar{v}_{s,0}(d\rho/dr)_{\text{eff},0}\omega^2 r_0}, \tag{23}$$

where

$$\left(\frac{d\rho}{dr}\right)_{\text{eff}} = \left(\frac{d\rho}{dr}\right) + \frac{\rho_0}{\bar{v}_{s,0}}\left(\frac{d\bar{v}_s}{dr}\right). \tag{24}$$

The uniqueness of these authors' treatment is that they were able to recast Eq. (24) such that the effective density gradient, $(d\rho/dr)_{\text{eff}}$, could be obtained experimentally. Their treatment is based on the experimental evidence obtained for DNA that the partial specific volume of the solvated species, $\bar{v}_s$, depends only upon the pressure and the activity of water at band center at atmospheric pressure, a_1^0. They were then able to obtain expressions for the buoyant density as a function of pressure and for the effective density gradient.

$$\rho_0^0 = \frac{1}{v_{s,0}^0}[1 - \psi P] \tag{25}$$

and

$$\left(\frac{d\rho}{dr}\right)_{\text{eff}} = \left[\frac{1}{\beta^0} + \psi\rho^{0^2}\right](1 - \alpha)\omega^2 r. \tag{26}$$

where

$$\alpha = \left(\frac{\partial\rho_0^0}{\partial a_1^0}\right)_P \left(\frac{da_1}{d\rho^0}\right) \tag{27}$$

and

$$\psi = \frac{\kappa - \kappa_s}{1 - \alpha}. \tag{28}$$

The term κ_s is a newly defined quantity, the apparent compressibility of the solvated polymer.

It is pleasing to note that the effective density gradient can be expressed in a readily interpretable form as:

$$\left(\frac{d\rho}{dr}\right)_{\text{eff}} = \left(\frac{d\rho^0}{dr}\right) - \left(\frac{\partial\rho_s^0}{\partial r}\right)_P + (\kappa - \kappa_s)\rho_0^0\omega^2 r. \tag{29}$$

The gradient which is effective in determining the polymer distribution is the composition density gradient plus the effective compression gradient less the density gradient associated with solvation changes of the polymer.

Equation (25) was verified by Hearst et al. (1961) who determined by additions of varying amounts of immiscible oil to the top of the gradient column and by variation of the speed that the buoyant density, ρ_0^0, did indeed vary linearly with pressure for tobacco mosaic virus and T-4 DNA as shown in Fig. 12. The differences, $\Delta\rho = \rho_0^0 - \rho_e^0$, are plotted as functions of the pressure. These experiments directly yield values of ψ. Thus far, no comparable pressure studies have been attempted for proteins. The term α is found by determining the buoyant density of the polymer in several different solvents in which the water activity varies depending on the molality of the salt solution required to band the polymer.

Fujita (1962) similarly treated the case in which there is interaction of the polymer with one of the solutes. He derived an expression for the square of the standard deviation involving the term $(d\Gamma'/dr)$ but did not

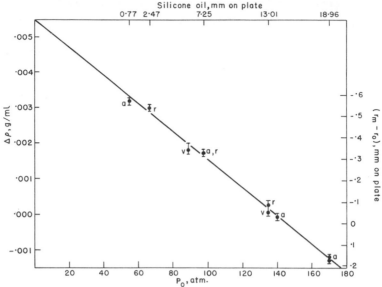

FIG. 12. Buoyant density increments for T-4 DNA in a CsCl solution at various pressures, $\rho_e^0 = 1.699$ g/ml at 25°C at 44,770 rpm. a, addition of oil; r, removal of oil; v, points obtained at 31,410 rpm and 39,460 rpm. The coordinates at the top and right-hand side indicate the original data. Slope $= -3.94 \times 10^{-5}$ g/ml/atm. The maximum error intervals are indicated. (Hearst $et\ al.,$ 1961.)

indicate how this could be determined. His expressions for σ^2 and ρ_0 reduce to those of Meselson, Stahl and Vinograd for the limiting case in which $(d\Gamma'/dr) = 0$.

Ifft and Vinograd (1966) have considered the case in which there is interaction of the polymer with both components of the solvent. All proteins bind extensive amounts of water and some proteins, such as serum mercaptalbumins, are known to bind significant numbers of anions. These workers invoked the concept of electroneutrality of the polymer to yield the following expression for the buoyant density:

$$\rho_0 = \frac{1 + z_2 + \Gamma'_*}{\bar{v}_3 + z_2\bar{v}_2 + \Gamma'_*\bar{v}_1}, \tag{30}$$

where z_2 is the grams of salt bound per gram of protein and $\bar{v}_2$ is the partial specific volume of the bound salt. The meaning of Γ'_* will be discussed in Section IV.

III. EXPERIMENTAL

The major portion of the discussion of experimental techniques will center around the use of the Spinco Model E analytical ultracentrifuge.

The reasons for this selection are outlined in the Introduction. Clearly, the discussion of the selection of the solvent, angular velocity, initial solution density, solution preparation, etc., can be directly extended to experiments in preparative centrifuges. A brief discussion of techniques pertinent to preparative ultracentrifugation concludes this section.

A. Analytical Ultracentrifugation

1. Selection of Experimental Conditions

Choice of solvent

There are no pure liquids of sufficient density in which proteins are soluble which will band proteins. Thus, the selection of a suitable two-component solvent is required. Because of solubility considerations and because it is generally desired to keep the ionic strength as low as possible, the selection of a 1:1 electrolyte is generally made. That this is not a very restrictive requirement is seen by the fact that the Merck Index (1960) lists thirty-four 1:1 electrolytes which have a density greater than 1.30 at saturation. This encompasses a wide selection of compounds which include all of the alkali metal ions and all of the halide ions. Care must be exercised that ρ_e not approach the saturated density too closely or crystallization may occur at the bottom of the cell during the experiment.

The solubility and stability of the protein in the salt solution of buoyant composition must be considered. The solubility can be checked in a straightforward bench-top experiment. The state of the protein in the concentrated salt solution can be examined via a physical chemical technique such as optical rotatory dispersion (Todd, 1960).

Fortunately, availability of most of these salts is not a problem. Sources of highly purified cesium and rubidium salts, of which the author is aware, are:

> American Potash and Chemical Corp., Los Angeles.
> Gallard–Schlesinger Corp., Carle Place, N.Y.
> Harshaw Chemical Co., Cleveland, Ohio.
> K & K Laboratories, Inc., Plainview, N.Y.
> Penn Rare Metals, Inc., Revere, Pennsylvania.
> Pierce Chemical Co., Rockford, Illinois.
> S. H. Cohen Associates, Yonkers, N.Y.

Preparative methods for salts not commercially available and purification techniques for salts of lower purity are given by Hearst and Vinograd (1961b), Szybalski (1967) and Wright *et al.* (1966). Other alkali metal and ammonium salts can be obtained in satisfactory purity from standard chemical sources.

Finally, a considerable number of ancillary questions should be weighed. Has a set of β-data been determined for this salt? Has a density versus refractive index relation been established? If the answers to these questions are no, a somewhat longer experimental and analysis time will be required. If resolution is to be an important factor in the experiment, a salt with a large β-value should be selected. If ultraviolet optics are to be used, the optical density of the saturated salt solution must not be above about 0.05 o.d. for a 1-cm cell at the wavelength of interest.

Selection of photographic system

Ultraviolet optics have been used almost exclusively in the study of DNA in density gradients because of the extremely high extinction coefficient of nucleic acids at 260 mμ. This high absorption results in a 100-fold increase in sensitivity of the ultraviolet optical system over schlieren optics. Proteins have a much lower extinction coefficient at the maximum of their absorbance curves at 280 mμ which results in only a two-fold increase in sensitivity. To offset this slight advantage, there are two major problems when absorption optics are used to record the equilibrium distributions of proteins. As is evident from Fig. 19 the location of the center of a rather broad concentration curve is extremely difficult and uncertainties of ± 0.005 g/ml in measured buoyant densities were recorded by Cox and Schumaker (1961). If a molecular weight is to be computed, the area under the curve must be integrated. Due to the relatively broad bands exhibited by proteins, it has been found that interpolation between the ends of the curve is completely unreliable and an experimental baseline must be provided. This requires a second run if absorption optics are employed. The use of schlieren optics provides values of ρ_0^0 accurate to ± 0.001 g/ml and a baseline can be simultaneously recorded on the same photograph as the solution refractive index gradient distribution.

Because of the distortion of the interference fringes at the high velocities required in density gradient centrifugation, little work has been done with interference optics to date.

Selection of angular velocity

In general, the highest centrifuge speed possible should be selected. This will afford the maximum gradient and therefore the narrowest band. Because the first-order theory has been developed for a constant density gradient through the band, the narrower the band the closer the experiment will fit the theory. The maximum centrifugal field will also be required even to fit the protein band within the liquid column for smaller

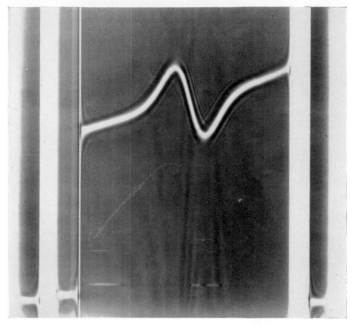

FIG. 13. Equilibrium schlieren photograph of 0·1% BMA in CsCl of $\rho_e = 1·279$ g/ml, pH $= 5·08$, acetate buffer ($\mu = 0·01$) at 56,100 rpm and 25°C. Single sector, 4°, Kel-F centerpiece. (Williams and Ifft, in preparation.)

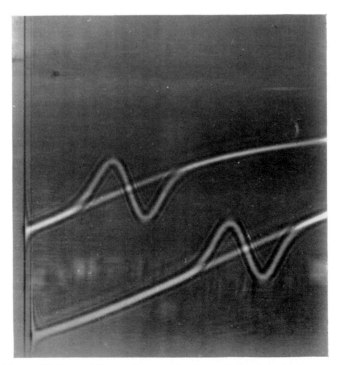

FIG. 16. Equilibrium schlieren photograph of 0·05% BMA in Cs_2SO_4 solutions of $\rho_{e,1} = 1·190$ and $\rho_{e,2} = 1·301$, pH 5·5, acetate buffer ($\mu = 0·01$) and baselines at 56,100 rpm and 25°C. Two double-sector, capillary-type synthetic boundary, $2\frac{1}{2}°$, filled-Epon centerpieces. (Ifft and Vinograd, 1966.)

proteins such as ribonuclease because of the approximate relation, $\omega \alpha 1/\sqrt{\sigma}$.

Equation (13) may be rewritten as

$$\omega^4 = \frac{RT\rho_0}{Mr_0^2\sigma^2}\beta_0^0.$$ (31)

Let us consider a typical experiment for a protein of molecular weight 10^5. If we choose to restrict the width of the band to a $\sigma = 0.1$ cm and assume a $\rho_0 = 1.3$ g/ml, $r_0 = 6.6$ and $T = 298.2°$K, we can derive the order of magnitude relation that

$$\omega^4 = 7·4 \times 10^5\beta_0^0.$$ (32)

In order to restrict the band width of such a system to 1 mm in a CsCl gradient, the maximum velocity of 55,000–60,000 for an aluminum alloy, An-D rotor is required. If β is doubled, as for comparable densities for KBr and RbCl, ω goes up by a factor of $\sqrt[4]{2}$ and the new An-H titanium rotor is needed. This rotor is relatively expensive but it can be run at 67,000 rpm.

Centerpiece selection

The choice of which one of numerous types of available centerpieces depends in large measure on what information is sought. If only the buoyant density is required, a single sector cell will suffice for most purposes. It is the experience of the author's laboratory that buoyant densities accurate to within ±0.001 g/ml can be obtained using only the schlieren image provided by the solution in a single sector centerpiece provided that σ does not exceed 1 mm. See Fig. 13.

If the amount of material available is limited, 2° sector centerpieces may be used instead of 4 and the exposure times doubled. The An-D rotor is designed to accept cells with $1\frac{1}{2}$-, 3-, 6-, 12-mm thick centerpieces, Generally the 12-mm centerpieces are used to provide the lowest possible concentration. Centerpieces of 18- and 30-mm thicknesses can be employed for even lower concentrations. However, then the An-E rotor must be used which has a maximum rated speed of 50,740 rpm.

A choice of materials is also available: aluminum alloy, Kel–F, a trifluorochloroethylene polymer, and filled-Epon, aluminum particles imbedded in an epoxy resin. Spinco recommends that in general aluminum centerpieces be employed because they distort less and because they can be run at elevated temperatures. The Kel–F and filled-Epon centerpieces, however, do not require gaskets and they are not subject to corrosion. Scratching of an aluminum centerpiece in the presence of a concentrated salt solution is neither good for the centerpiece nor the material under investigation.

On the other hand, if an especially wide band is encountered or if the distribution is to be analyzed for the molecular weight of the protein, experience has shown that an experimentally determined baseline is an absolute requirement. This correct baseline can only result if the column lengths are identical, the initial solution densities are identical and the protein concentration is low enough so that the CsCl distribution in the protein solution is not affected. Ifft and Vinograd (1961) investigated a variety of methods to provide this baseline. A cell with a 4°, aluminum, centerpiece was filled with a CsCl solution and weighed. After centrifuging this solution to equilibrium, the contents of the cell were withdrawn and an exactly equal mass of protein–CsCl solution was inserted. Exact superposition of the two equilibrium photographs was found to be difficult. The second method investigated was the use of a regular double-sector, filled-Epon centerpiece. Spinco supplies these centerpieces with $2\frac{1}{2}°$ sectors (see Fig. 14b). Identical volumes of solvent and solution were

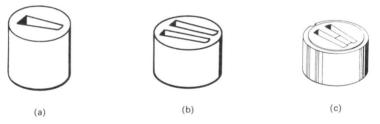

(a)   (b)   (c)

Fig. 14. Basic set of centerpieces useful in density gradient work. (a) Single sector, 12 mm, 4°, (b) Double sector, 12 mm, $2\frac{1}{2}°$. (c) Double sector, capillary-type synthetic boundary, 12 mm, $2\frac{1}{2}°$. Reprinted from Spinco's Instruction Manual. E-IM-3.

injected into the two sides of the centerpiece with a precision syringe and a micrometer head. This procedure was found to yield identical column lengths in only about one-third of the trials in which it was used due to a number of experimental difficulties.

The most successful procedure utilized a double-sector, capillary-type synthetic boundary centerpiece available from Spinco. This centerpiece, as can be seen in Fig. 14c, has two, $2\frac{1}{2}°$ sectors connected by two fine capillary grooves, about 0.001×0.001 in. If one sector is filled to a point below the middle capillary with solution and the other sector is nearly filled with solvent, solvent will transfer through this capillary at about 7000 rpm until the column lengths are absolutely identical. The upper capillary is to provide for air-pressure equilibration. The disadvantage of this type of experiment is that a small amount of protein generally transfers into the baseline solution by diffusion.

Several new cell designs have been described by Ende (1964). The simplest is presented in Fig. 15. Construction begins by machining two small holes A on the sides of a double-sector, filled-Epon centerpiece.

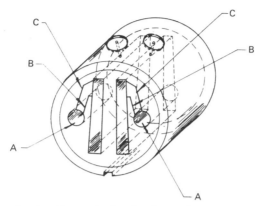

FIG. 15. New centerpiece for conducting density gradient experiments to provide identical column lengths of solution and solvent. See text for description of construction and operation. (Ende, 1964.)

These are connected to the sectors by machined grooves B and C which provide identical column heights and pressure equalization respectively. The cell is filled with solvent and solution on the two sides above groove B. The centrifugal field forces the excess fluid into reservoir A. The principal problem appears to be precise machining. Two modifications of this basic design are presented.

Choice of rotor and windows

Spinco provides five rotors which might be used for these experiments. Their properties are summarized in Table IV.

TABLE IV. SPECIFICATIONS FOR SPINCO ANALYTICAL ROTORS

	Rotor type				
	An-D	An-E	An-F	An-G	An-H
Maximum speed	59,780	50,740	52,640	42,040	68,000
Maximum number of cells	2	2	4	5	2
Centerpiece thickness accepted	1½ to 12	18 to 30	1½ to 12	1½ to 12	1½ to 12

The standard rotor which has been used most frequently is the An-D. It accepts a 12-mm cell and a counterbalance or two cells and until recently had the highest velocity rating. Its price of $640 vs. $2520 for the titanium rotor may influence its continued use.

As more workers have come to realize the value of their time and machine time, the practice of two cell experiments has come into more frequent usage (see Fig. 16). The two cells are filled with the appropriate volumes of the solutions of interest to yield a difference in mass between the cells of less than 0.5 g. One of the cells is assembled with a −1 wedge in the upper window holder to depress one of the images on the plate relative to the other. The reference hole and balance hole in the rotor are unplugged to yield an absolute distance from the center of rotation.

Szybalski (1967) has described an elegant method for doing four-cell experiments using the An-F rotor and a variety of flat and side-wedge windows. His method applies exclusively to the use of ultraviolet optics. Conceivably, four non-overlapping schlieren images could be formed using the An-F rotor by utilizing the combinations of windows given in Table V.

TABLE V. POSSIBLE COMBINATIONS OF
WEDGE WINDOWS FOR USE IN
THE An-F ROTOR

Cell no.	Top window	Bottom window
1	Flat	Flat
2	−1°	Flat
3	−2°	Flat
4	−2°	−1°

It is unlikely that six-cell experiments will be performed in the near future with the An-G rotor until satisfactory procedures have been developed for using the interference optical system at these high speeds and large gradients.

It will frequently be necessary to use negative wedge windows to record the image on the plate even when only one cell is run. The refractive gradients generated by some of the heavier salts such as CsI and Cs_2SO_4 require combinations of wedge windows on the top and bottom of the cell to yield −3° or even −4° depression. Figure 13 was obtained with two flat windows. The position of the gradient curve on the photograph can be compared with that of Fig. 17 in which a −1° window was used. Great care must be exercised to keep each wedge window in the window-holder that was machined to fit it and that the notches on the window and on the holder match up exactly when the cell is assembled.

Selection of initial density

This is a relatively simple problem for pure proteins as discussed in the introductory portion of the theoretical section. A density of 1.29 will probably yield a recognizable schlieren pattern because the buoyant densities of all pure, soluble proteins in CsCl have fallen in the range

1.24–1.34. From this initial experiment, an approximate value for the buoyant density can be obtained. Then a second run at a modified density should yield a suitable band near the center of the cell.

As mentioned earlier, any bound lipid, saccharide or nucleotide will alter the above prediction.

Other parameters

Most experiments are conducted at 25° C because this is the temperature for which most β-values and refractive index relations have been established. However, Hu *et al.* (1962) did their work and their gradient calculations at 20° C. For the enzyme chemist, the density gradient technique would be far more interesting if experiments could be performed at temperatures near 0° C. The Model E is equipped to operate at 0° C provided that a slow stream of dry N_2 is passed over the lower side of the chamber collimating lens to prevent fogging. Density and molality data are available down to this temperature but activity coefficients or osmotic coefficients are not generally available.

The concentration of protein to use depends in part on the solvent selected. In CsCl, a convenient concentration has been 0.1% (w/v). This provides a sharp schlieren image at a bar angle of 55° to provide reasonably good images for analysis of the plates. This concentration can be decreased to 0.03% for runs in salt solutions having particularly high density gradients such as CsI and Cs_2SO_4.

The solution, of course, must be buffered. This generally will be at the isoelectric point of the protein unless a buoyant density titration is contemplated. The buffer concentration should be as low as possible to avoid distortion of the gradient and minimize competitive anion binding effects if they exist. A $\mu = 0.01$ is generally satisfactory to hold the pH where it is desired provided the pH is close to the pK_a of the buffer.

2. Run Preparations

Preparation of solutions

In general, the required solution can be described as follows:

Volume: ~ 2 ml.
Protein concentration: 0.1%.
Buffer concentration: $\mu = 0.01$.
pH: Isoelectric point of the protein in this concentrated salt solution.
Density: estimated ρ_0^0.

A volume of 2 ml is suggested to provide an adequate amount of solution to repeat the run in case a leak occurs, to provide for losses in the syringe, the volume required for refractometry, etc. Single-sector runs require about 0.7 ml and double-sector cells about 0.4 ml to provide a maximum gradient column.

A relatively simple way to proceed is to prepare the following stock solutions.

(a) A nearly saturated salt solution. This can be prepared following the data in the Merck Index (1960) or it can be empirically produced by continued addition of the salt to 5 or 10 ml of water. It is convenient to prepare about 20 ml of the saturated solution if a series of 5–10 runs is planned. The optical density for a 1-cm path length at 280 mμ should be determined in the event absorption studies are ever undertaken with that solution and as a rough measure of organic impurities. The refractive index and pH should also be recorded. If a refractive index-density relation is not known, the density also should be measured. The solution can be conveniently stored in a small capped plastic bottle and kept at room temperature. We have found virtually no trouble arising from bacterial growth in these concentrated salt solutions.

(b) A 1% protein solution. This solution should be that obtained directly from an ion-exchange column or the solution obtained after exhaustive dialysis vs. distilled water. A tremendous advantage in density gradient work is that a knowledge of the actual protein concentration present in the salt solution is not required. Thus this solution need have only a nominal 1% concentration.

Because of the effort and expense involved in a density gradient experiment, there should be a reasonable expectation that the protein sample is quite pure and homogeneous. A few brief but excellent accounts of methods of preparing or obtaining a number of typical, well-characterized proteins are given by Sobotka and Trurnit (1961). Thus a preliminary sedimentation velocity experiment to demonstrate a symmetrical sedimenting boundary should be the minimum criterion before a density gradient experiment is attempted. If electrophoretic or solubility criteria can also be demonstrated, so much the better.

(c) Buffer solution of correct pH, $\mu = 0.1$. Liter quantities can be prepared according to a favorite set of empirical tables or standard analytical calculations, adjusted with either the anion or weak acid to the exact pH desired and stored in the cold room. Standard buffers frequently employed are given in volume 2 of this series by Magdoff (1960) and Kenchington (1960) and in volume 1 by Polson (1960).

Mention should be made of the new set of buffers recently developed for use in the range pH 6.8 (Good et al., 1966). These buffers, designated as Good's buffers, are available from Calbiochem, Los Angeles.

Now the relative volumes of water and concentrated salt solution are computed using one of the two relationships given below. If the final density needs to be exactly right upon mixing the two solutions, the non-additive relationship should be employed. However, only one weight fraction-density relationship has been published (Eq. (7)). Also, unless the experimenter is quite skillful in manipulating serological pipettes and micropipettes, the final density will require some adjustment in any case. Thus the use of Eq. (33) is recommended. The solution thus prepared will probably require small additions of concentrated salt solution or water to achieve the desired density.

$$\text{Additive volume relationship:} \quad V_1/V_2 = -\frac{\rho - \rho_2}{\rho - \rho_1}. \tag{33}$$

$$\text{Non-additive volume relationship:} \quad V_1/V_2 = -\rho_2/\rho_1 \frac{(F - F_2)}{(F - F_1)}. \tag{34}$$

Symbols without subscripts refer to the desired solution. The subscripts 1 and 2 refer to concentrated salt solution and water in either order. The symbol F is the weight fraction of salt.

A reasonable value for, say, V_2 is assumed for the amount of stock salt solution which is required. Then a volume of water, V_1 is calculated. Assuming additivity, the final volume is computed as $V_1 + V_2$. The volumes of protein solution and buffer should be one-tenth of this total volume. Because the protein and buffer solutions are considered to be pure water, the actual volume of water to add is V_1 diminished by these two volumes.

These four volumes are carefully measured with 0.1, 0.2 and 1.0 ml Mohr or serological measuring pipettes calibrated to read to 0.01 ml and the solution thoroughly mixed. Small vials with plastic caps provide convenient storage vessels.

Measurement of density

A number of methods are available. The most convenient is refracto-metry. Linear relationships between refractive index and density have been established experimentally or derived from data in the International Critical Tables (1928) by various workers for a number of salts. These extremely useful relations have been summarized in Table VI. The data yield densities valid at 25.0°C from refractive indices measured at the sodium D line at the same temperature. Most of the equations are accurate to within ±0.001 g/ml. The original references should be consulted for the actual uncertainties.

Use of the equations to yield densities accurate to 0.001 density units requires the measurement of the refractive index to ±0.0001. The Abbé type refractometer is best suited for this purpose because this is precisely

TABLE VI. DENSITY VERSUS REFRACTIVE INDEX FOR AQUEOUS SALT SOLUTIONS

$$\rho^{25°} = a(n^{25°}) - b$$

Salt	Coefficients of equation a	b	Density range	Reference
Cs acetate	10.7527	13.4247	1.80–2.05	Hearst (1961)
CsBr	9.9667	12.2876	1.25–1.35	Ifft and Vinograd (1966)
CsCl	10.2402	12.6483	1.10–1.38	Bruner and Vinograd (1965)
	10.8601	13.4974	1.37–1.90	Ifft, Voet and Vinograd (1961)
Cs formate	13.7363	17.4286	1.72–1.82	Hearst (1961)
CsI	8.8757	10.8381	1.20–1.55	Ifft and Vinograd (1966)
Cs_2SeO_4	12.0919	15.1717	1.38–2.00	Hearst (1961)
Cs_2SO_4	12.1200	15.1662	1.15–1.40	Ifft and Vinograd (1966)
	13.6986	17.3233	1.40–1.70	Hearst (1961)
KBr	6.4786	7.6431	1.10–1.35	Ifft and Vinograd (1966)
NaCl	4.23061	4.64125	1.00–1.19	Vinograd (unpublished)
RbBr	9.1750	11.2410	1.15–1.65	Ifft and Vinograd (1966)

the limit of accuracy for it and because only one drop of solution is required. Szybalski (1967) has referred to the possibility of ordering a special narrow gap between prisms when a new instrument is obtained. We have examined the two most popular instruments on the market and have found that the Bausch and Lomb Abbé 3L refractometer yields the highest accuracy in determining the fourth decimal place.

A constant temperature bath must be used to circulate water at $25.0 \pm 0.1°C$ through the prisms. The temperature should be read from the thermometer at the circulating outlet on the instrument after this thermometer has been calibrated.

The refractometer should be periodically calibrated and either adjusted internally to yield exactly correct values or else the correction value posted prominently near or on the instrument. The test piece supplied with the instrument provides the greatest reliability and accuracy for this purpose. A quick check with distilled water is frequently useful.

In addition to the calibration correction, a small correction should be applied to each measurement of the protein–salt solutions to account for the contribution of the protein to the refractive index. This contribution can be approximated with sufficient accuracy by $(dn/dc) \cdot c$ where (dn/dc) is the specific refractive increment of the protein at this temperature, pH and ionic strength and c is the protein concentration in g/dl. Perlman and Longsworth's (1948) data indicate that this correction can be given as $0.0019 \cdot c$ in all cases. This value must be subtracted after the correct refractive index for the protein solution is obtained.

If large quantities of solution are available and extreme precision is required, a differential refractometer such as the Phoenix instrument can be put to good advantage (Ludlum and Warner, 1965).

Another popular method of determining densities of small quantities of liquids is by using micropipettes. Hearst and Vinograd (1961b) have described the use of a 300-μl micropipette to determine densities accurate to ±0.001. The pipette is calibrated with distilled water and then filled with the solution and reweighed. An ordinary analytical balance having an 0.1-mg sensitivity is satisfactory. No temperature correction is required if room temperature is within a few degrees of 25°C. More recently, Vinograd *et al.* (1965) have described the use of a 100-μl pipette and a micro-balance to obtain densities to ±0.001 d.u.

If large volumes of salt solution are available, the standard procedures utilizing pycnometers (Ludlum and Warner, 1965) or a Westphal balance (Ifft and Vinograd, 1966) can be used.

Measurement of pH

An accurate measure of the pH of the protein solution being examined is almost always required, whether or not a buoyant density titration is performed. If sufficient protein and salt are available, 3–5 ml of solution can be prepared. An ordinary glass electrode and calomel reference electrode can therefore be used. However, at these low buffer capacities, it is important to check the pH of the protein solution at the conclusion of the run to make sure that the pH has not drifted. Because only about 0.6 ml of solution can be recovered in a single-sector experiment and 0.35 ml in a double-sector experiment, a different pH measurement system is required. Excellent results (±0.005 pH unit) have been obtained in the author's laboratory using the Beckman combination electrode No. 39030 in conjunction with the Beckman research pH meter, model 1019, providing that the electrode is connected to a reservoir of saturated KCl to provide hydrostatic pressure on the reference junction. If the solution is transferred from the centrifuge cell to a small test-tube just large enough to admit the 5-mm bore of the electrode, the pH of only a few tenths of a milliliter of solution can be obtained. Several other companies have similar combination pH electrodes available.

3. Conduct of the Run

Detailed instructions for the operation of the Model E ultracentrifuge are provided in the new Instruction Manual E-IM-3. This is an excellent manual and owners of Model E's which were purchased before the manual was issued would find this to be a wise $18 investment. Therefore, only a few points which are particularly troublesome or have special relevance to the exacting requirements of the extended times required for equilibrium experiments will be covered in this section.

A sample protocol sheet is included in Appendix A. It is most helpful to have available a supply of sheets similar to this for the conduct of the run, for subsequent analysis and to provide a permanent record of the primary data of each experiment.

Cell assembly and filling

The cell should be assembled giving careful thought to requirements of a particular centerpiece or wedge windows. The new cell torque wrench (Spinco part No. 327119) is an extremely convenient tool to provide a uniform and reproducible torque each time. A torque of 120 in.-lb is generally sufficient to prevent leaks. If a leak should occur, an increase in torque by 5 in.-lb at a time will provide a tight cell.

All needles used in filling should have their tips ground flat to avoid scratching the centerpiece. No particular filling instructions are required unless a double sector experiment is to be made in which case comments made in the above section entitled "Centerpiece selection" are appropriate.

The standard procedure of balancing the cell and counter-balance to within 0.0 to +0.5 g is especially pertinent here because of the variety of centerpieces and the higher densities of the solutions.

Because of the concern for an accurately determined initial density, the use of two filling-hole, polyethylene gaskets and acceleration to about 2000 rpm to seat these gaskets before full vacuum is achieved is recommended.

If the pressure corrections discussed in Section II of this chapter are to be applied to the buoyant density, the length of the liquid column and band positions different from the isoconcentration position within reasonable limits are acceptable. If Eq. (9) will not be applied, every effort must be expended to band the protein at r_e and in a series of experiments, to keep the column lengths identical. It should be noted that virtually none of the buoyant densities reported in the literature have been corrected for pressure.

Calibration of the thermistor

As in all centrifuge work, an accurate knowledge of the temperature is a prerequisite. Thermistors are known to age considerably during their first 100 hr of operation. It is our experience that ageing continues for an indefinite period beyond this. Every 3 or 4 months, the thermistor should be recalibrated and the resulting RTIC (Rotor Temperature Indicating and Control) potentiometer reading plotted against temperature. The dated plot should be kept near the ultracentrifuge at all times for ready reference.

Derating of the rotor

A careful log should be kept of hours each rotor is run at full speed. Spinco calls for derating the maximum speed by 10% for every 1000 runs or after 1000 hr at maximum speed, whichever comes first. Our experience indicates that operation of the rotor for the extended periods of time that are required in this type of work stresses the rotor far less than an equivalent number of hours put on the rotor in velocity work. Because of the careful engineering design of the armored rotor chamber, the decision is primarily an economic one. How many rotor lifetimes are equivalent to the cost of rebuilding the chamber? The decision is important because the cost of rotor depreciation for equilibrium density gradient runs based on a 1000-hr lifetime is $16/run.

Of course, the maximum rated speed stamped on the rotor support ring must never be exceeded.

Time to reach equilibrium

An estimate of this quantity is available from Eq. (14). Because this equation was derived assuming that the salt gradient had already been established, the length of time the experiment is conducted to achieve sensible equilibrium must always be established experimentally.

An irregularly shaped band is formed after 6 or 7 hr of operation for a typical protein. Depending upon how close ρ_e is to ρ_0, a symmetrical band is formed at about 8–9 hr at essentially the correct banding position. However, this band narrows with a corresponding increase in height of the inflection points until no further change is noted after 15 hr. In general, we found that for most proteins in most salt solutions, 24 hr is sufficient time to approach closely enough to equilibrium that no further changes in band shape or position can be detected by continued operation.

Photography

The schlieren optical system will probably be used. A variety of plates and films are available. Spectroscopic plates, Types I-N, II-G and ID-2, might be selected for certain applications. However, because of the cost and stability factors, Kodak's metallographic plates are frequently employed. They may be obtained directly from Spinco or ordered through most local camera stores. The exposure time will vary with the age of the lamp. Generally a 3–5-sec exposure will produce an image with satisfactory contrast for analysis.

Recently, 2×10 cut film has become more readily available. Also it can be cut in the darkroom from 8×10 sheets. It can be used in the 2×10 plate-holder in conjunction with the adaptor. It does not provide as accurate a photometric record for measurement as the glass plates but because it reduces photography costs by a factor of 5, it should be used where a large number of exploratory runs are required.

The schlieren lightsource is a high-voltage and high internal-pressure lamp produced by General Electric Co. and designated as the Type A-H6 lamp. An adequate water supply, 3 quarts/min, must be connected to the lightsource housing which will prevent power from being supplied to the lamp unless the water flow is started and is maintained. Otherwise the lamp will instantly disintegrate. Instructions are supplied with each lamp as to how to distribute the mercury prior to the initial start. A section of lamp insulation (available from Spinco) should be slipped over the high-voltage end of the lamp just up to the metal–glass juncture. *Gentle* tapping on the housing may help if the lamp fails to start. Spinco's instructions should be followed with regard to rotation of the lamp and adjustment of the slit jaws. The lamp should be left on only when viewing or photographing the image unless the lamp will be turned on again within an hour. Deposition and corrosion require occasional replacement of the water jacket. A supply of A-H6 lamps, pyrex water jackets, silicone gaskets, brass inserts and lamp insulation should be kept on hand.

Determination of the mean angular velocity

An average value for ω can be obtained by making odometer readings at two measured times. Multiplication of observed readings by 6400 converts them to revolutions if the drive selector is in the HI position. Conversion of these values to radians and division by the elapsed time in seconds yields the angular velocity in the desired units.

Alternatively, a nearly exact value for ω can be obtained directly from the gearbox setting. If these values are employed, the table given in Appendix B will prove useful in providing an accurate and rapid conversion from any of the nominal angular velocity settings in rpm to ω^2 values in (radians/sec)^2.

Common reasons for premature shutdown

The ultracentrifuge is a sophisticated instrument. Successful operation over extended periods of time requires a high level of performance from a number of components. Ultracentrifugers are fortunate that the Spinco Model E ultracentrifuge is an extremely well-designed instrument and trouble incidence is at a pleasingly low level. Among the following items, none can be listed as the predominant problem because with proper maintenance by a qualified field engineer and by the operator himself, every one of these problems can be eliminated.

(a) *Loss of vacuum*

The Vacuum Indicator should be checked routinely. If it indicates a rapid rise in pressure, the run should be immediately stopped. Vacuum losses frequently arise from cell leaks which are caused by faulty cell assembly or inadequate torquing. Vacuum can also be lost by an improperly greased O-ring at the top of the chamber or, of course, faulty functioning of either of the vacuum pumps.

(b) *Blurred images*

If blurred horizontal streaks appear in the image, the most probable source of difficulty is oil on the lower collimating lens. If the drive has seen extensive use, the bearing may be worn and oil may leak down the shaft, strike the rotor and be dispersed as a fine mist. Care should be taken that the bottom of the chamber be kept free of oil and that the lenses are wiped clean prior to each run. A Kimwipe moistened with acetone followed by a dry Kimwipe will do an adequate job for schlieren optical work. Care should also be taken that no more than the minimum amount of silicone oil be applied to the pool of mercury. An excess of oil can be volatilized by the heating wire. A minute nick in the contact needle will cause the mercury cup to be promptly emptied. The needle should be periodically inspected with a small magnifying-glass to be certain it stays sharp. The needle height should occasionally be checked and correctly positioned for that particular thermistor assembly.

(c) *Loss of temperature control*

This can often be traced to a faulty contact between needle and mercury and can be prevented by maintenance described immediately above. Occasional electronic difficulties arise with the RTIC system.

4. Conclusion of the Run

Development of the plates

Successfully developed plates must always be made before the centrifuge is shut down. If a box of plates has been accidentally exposed, the developer is not effective, the plate-holder drive was not in the "IN" mode during photography or any other slip in the photographic operation has occurred, the run can still be a complete success if the operator knows this while the run is still in progress and can act appropriately.

Because intensity measurments are not required from schlieren plates, there is considerably greater latitude in the development of these plates than in ultraviolet work. For instance, we routinely have obtained satisfactory results without any temperature control of the photographic solutions in a room which has a temperature of about $70 \pm 8°C$. Also only a modest amount of manual agitation is required in any one of the development steps. Plastic $4 \times 4 \times 12$-in. trays with covers provide convenient developing vessels. All stock solutions must be kept in dark bottles in a dark cupboard and discarded after 4–6 weeks.

The following development procedure for Kodak metallographic plates is recommended:

1. Remove plate from cassette and write run number and plate number across one end of the emulsion in pencil. This is done in the dark or under the appropriate safelight.
2. Place in Kodak D-11 developer for 5 min. Occasionally agitate.
3. Place in Kodak Indicator Stop Bath for 30 sec.
4. Place in Kodak General Purpose Hardening Fixer for 10 min. Lights may now be turned on.
5. Rinse in a tank of slowly moving cold water for 1 hr.
6. Rinse briefly in Kodak Photo-flo 200 solution.
7. Air dry.

All plates should be stored in numerical run order in a common file cabinet in $2\frac{1}{2} \times 10\frac{3}{4}$ in. envelopes (obtainable from Spinco). A suitable amount of information on the outside of the envelope such as run number, type of run and material examined provide a convenient and rapid access to any run of interest.

Re-examination of the solution

Immediately upon removal of the rotor from the chamber, the cell and cell contents should be visually examined for suspicious accumulations of dried salt or an unusual appearance of the solution. Then the solution should be thoroughly mixed by inversion of the cell 25 times or so or placing the cell on a roller assembly for several minutes. A clean syringe should then be employed to remove as much of the solution as possible. The refractive index, pH and any other property of the solution of interest should be promptly measured. If the solution properties have changed appreciably from the previous measurements, the initial solution which was not run must be examined again and a decision made as to what are the appropriate values.

5. Analysis

Most density gradient experiments require a rather thorough analysis to arrive at a buoyant density. The exception might be when the resolution of a sample is of prime interest and actual values of ρ_0 are not required.

Measurement of the plates

There are two primary techniques for the determination of positions on the plates. The first technique is using a two-dimensional comparator

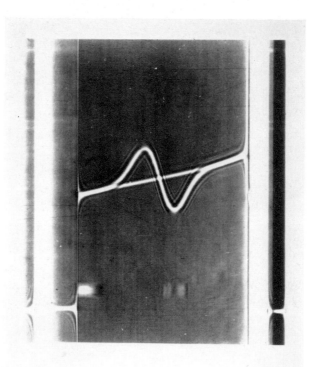

FIG. 17. Equilibrium schlieren photograph of 0·1% BMA in 2·59 molal CsCl, pH 5·5, acetate buffer ($\mu = 0·01$) and baseline at 56,100 rpm and 25°C. Double sector, capillary-type, synthetic boundary, $2\frac{1}{2}°$, filled-Epon centerpiece. (Ifft and Vinograd, 1962.)

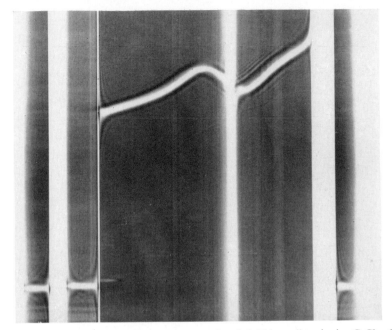

Fig. 21. Equilibrium schlieren photograph of 0·1% ovalbumin in CsCl of $\rho_e = 1\cdot278$ g/ml, pH$=4\cdot66$, acetate buffer ($\mu = 0\cdot01$) at 56,100 rpm and 25°C. Single sector, 4°, Kel-F centerpiece. (Zilius and Ifft, unpublished.)

such as the Gaertner Model M 2001-P (Gaertner Scientific Co., Chicago). The reading accuracy with such a comparator is 0.01 mm which corresponds to 5 microns in the cell. Despite this phenomenal precision, a comparator is not recommended for schlieren plate analyses because the instrument cannot be used for interpolation. This means that there is no way in which the inflection point can be detected in a single-sector experiment. Also, because of the nature of the optical system, the interference fringes obliterate portions of the refractive index gradient curves in photos obtained in a double-sector experiment with solution in one sector and solvent in the other. The nature of this difficulty can be realized by careful examination of the curves near the center and at the edges of the Gaussian band, shown in Fig. 17. If just the buoyant density is required the light area at the exact center of the band can be read with extreme precision. However, if a numerical integration of the area between the curves is needed for the determination of a molecular weight, the comparator then fails badly.

A simpler and less costly solution to the analysis problem is the use of a photographic enlarger. Most darkrooms are equipped with them for the usual printing process and they can be readily converted to the function of analyzing plates by the purchase or construction of a simple adaptor with a 2-in. groove in it. The author has found the Omega II-D and the Beseler 23C enlargers entirely satisfactory. They provide up to about 10- and 6-fold magnifications respectively. The image is projected onto a large sheet of white vellum paper in a darkened room. A sharply pointed 4H pencil is used to locate the lines of interest on the photograph. The eye is quite sensitive to the shading provided by the interference fringes and dots can be readily placed at the center of each schlieren line about 1 cm apart. Four or five dots will accurately locate the top and bottom reference edges and the top and bottom of the liquid column. Then the lights may be turned on and a ruler used to locate the latter four radial positions on the paper. A combination of french and boat curves will permit the smooth representation of the schlieren curves, including, by interpolation, those portions blurred by the intersection of the two lines.

If the buoyant density is all the information that is desired, the graphical work is now completed. If a molecular weight is required, the area between the two curves must be numerically integrated. The area should be divided into 5-mm wide intervals and the midpoint heights measured with a precision ruler beginning at band center and moving outward through the two bounded areas. The analysis is considerably speeded if all of this primary data is directly recorded from the tracing onto the tape of a printing calculator.

A necessary requirement for a successful analysis is that the areas under the two lobes of the biphasic curve be equal. This is required because the areas are directly proportional to the concentration at r_0. If the areas are divergent, the solution densities or column lengths were unequal or the baseline has been drawn incorrectly. This is not a sufficient condition, however. Arbitrary baselines have been drawn for single-sector experiments and adjusted until the areas were identical. The resulting log plots were not linear to 1σ indicating an incorrect baseline.

The precision of this method of integration as determined by repeated measurements of the same frame is about 1%. An independent check is to compare the ratio of two areas obtained from two photos taken at two different bar angles to the ratio of the cotangents of those bar angles. This has also been reproduced to about 1%.

Calculations

A thorough analysis of a centrifuge plate requires that a distressingly large number of distances and densities be determined. We have found it extremely helpful to provide copies of the following tabular outline to all workers beginning density gradient experiments. It is developed specifically for salts such as CsCl, RbBr, RbCl and KBr for which β values, n vs. ρ relations and normalized isoconcentration points are available. The format developed in the Notation section and subsections I, II, III and V of the Calculations section should be of general use, however.

The outline is self-explanatory. The following numbered comments refer to the superscripts in Table VII.

[1]Subscript zero refers to band center; superscript zero refers to atmospheric pressure.

[2]Obtain from Fig. 5 or data given by Ifft *et al.* (1961).

[3]Subscript *e* refers to the properties of the initial solution.

[4]Obtain from Table I and Figs. 6 and 7.

[5]Values of isothermal compressibility coefficients which will serve with sufficient accuracy for most protein experiments are given in Table VIII. More exact values can be obtained from interpolation and extrapolation of the data of Pohl (1906) and Gibson's (1934, 1935) data for the compression of several salt solutions. More recently, Gucker *et al.* (1966) reported measurements of the adiabatic compressibility of aqueous solutions of NaCl and KCl at 25°C. It is hoped that these measurements will be extended to other salts and into a higher concentration range.

[6]The actual distance between the counterbalance reference edges should be determined with an optical comparator and used in place of the nominal value of 1.600 given by Spinco.

[7]The last two decimal places depend on the extent of rotor stretch which in turn depends upon the particular rotor and the speed. Schachman (1959) gives a discussion of methods of measurement of

Log No. ....   Name ....

Plate-frame ....   Date ....

Material   ....

NOTATION

I. Distances
   A. In cell in cm
      r_a: meniscus of solution
      r_b: bottom of solution
      r_e': limiting isoconcentration point
      $r_0^{(1)}$: band center
      r_e: isoconcentration point
      r_α: root-mean-square distance between band center and meniscus

   B. On tracing in cm
      R_T: top reference edge
      R_a: meniscus of solution
      R_0: band center
      R_b: bottom of solution
      R_B: bottom reference edge
   C. Related distance parameters

   $\dfrac{r_e - r_a^{(2)}}{r_e' - r_a}$: normalized isoconcentration point

   M.F.: magnification factor from cell to tracing

II. Densities and density gradient
   ρ_0: density at r_0, $g\,cm^{-3}$
   $\rho_e^{(3)}$: density at r_e
   $\Delta\rho = \rho_0 - \rho_e$
   $\beta^{(4)}$: density gradient proportionality constant
   $(d\rho/dr)_0$: density gradient at r_0, $g\,cm^{-4}$

III. Other experimental parameters
   ω: angular velocity, radians/sec
   T: absolute temperature
   σ: standard deviation of Gaussian distribution, cm
   $\kappa^{(5)}$: isothermal compressibility coefficient of salt solution of density ρ_e

CALCULATIONS

I. Absolute temperature: 298.2°K
II. Angular velocity
   Odometer readings: Elapsed time:
   at t_2: $t_2 - t_1 = (1440 \cdot x\ \text{days} + 60 \cdot y\ \text{hr} + z\ \text{min})\ 60\ \text{sec}$
   at t_1: −   ....
   Δ odometer =. $= (1440 \cdot \quad + 60 \cdot \quad + \quad)\ 60\ \text{sec}$
   Total rev $= 6400 \times \Delta$ odometer
                $= ....$ rev $= ....$ sec

   $\omega^2 = \left(\dfrac{\text{total rev} \times 2\pi}{t_2 - t_1(\text{sec})}\right)^2 = \left(\dfrac{\quad \times 6.283}{\quad}\right)^2 = \quad \text{sec}^{-2}$

III. Distances

   $\text{M.F.} = \dfrac{R_B - R_T}{\text{Distance between reference edges}} = \dfrac{}{1.600^{(6)}} =$

197

$$r_a = 5.7^{(7)} \quad + \frac{(R_a - R_T)}{MF} = 5.7 \quad +(\ldots\ldots\ldots) = \qquad \text{cm}$$

$$r_b = 5.7 \quad + \frac{(R_b - R_T)}{MF} = 5.7 \quad +(\ldots\ldots\ldots) = \qquad \text{cm}$$

$$r_0 = 5.7 \quad + \frac{(R_0 - R_T)}{MF} = 5.7 \quad +(\ldots\ldots\ldots) = \qquad \text{cm}$$

$$r'_e = \sqrt{\{(r_a^2 + r_b^2)/2\}} = \qquad \text{cm}$$

$$r_\alpha = \sqrt{\{(r_0^2 + r_a^2)/2\}} = \qquad \text{cm}$$

$$\left(\frac{r_e - r_a}{r'_e - r_a}\right) =$$

$$r_e = r_a + \left(\frac{r_e - r_a}{r'_e - r_a}\right)(r'_e - r_a) = \qquad + \qquad = \qquad \text{cm}$$

$$\Delta r = r_0 - r_e = \qquad \text{cm}$$

IV. Density parameters

$$\rho_e^{0(8)} = \qquad n_D^{25°C} - \qquad = \qquad \text{gcm}^{-3}$$

$$\beta_e = \qquad \text{cgs}$$

$$(d\rho/dr)_e = \omega^2 r_e/\beta_e = \qquad = \qquad \text{gcm}^{-4}$$

$$\Delta\rho^0 = \Delta r(d\rho/dr)_e = \qquad = \qquad \text{gcm}^{-3}$$

$$\rho_0^0 = \rho_e^0 + \Delta\rho^0 = \qquad = \qquad \text{gcm}^{-3}$$

$$\beta_0 = \qquad \text{cgs}$$

V. Pressure corrections

$$\rho_\alpha = \rho_e^0 + (d\rho/dr)_e(r_\alpha - r_e) = \qquad \text{gcm}^{-3}$$

$$P_{r_0} = \frac{\rho_\alpha \times \omega^2}{2.026 \times 10^6}(r_0^2 - r_a^2) = \qquad \text{atm}$$

$$\rho_0^{(9)} = \rho_0^0(1 + \kappa P_{r_0}) = \qquad \text{gcm}^{-3}$$

$$(d\rho/dr)_{\text{phys},0} = 1/\beta_0 + \frac{\kappa\rho_0^{02}}{1.013 \times 10^6}(\omega^2 r_0) = \qquad \text{gcm}^{-4}$$

VI. Standard deviations

A. Schlieren optics. Numerically integrate schlieren curve from center of band outwards in 5-mm intervals.

B. Absorption optics. Measure height of concentration peak at an appropriate number of equal intervals.

In either A or B, using the following computation:

Plot $\ln c$ vs. $(r - r_0)^2$ using $(r - r_0)$ in cm on tracing.

$$\text{Slope} = \frac{(\ln c)_1 - (\ln c)_2}{(r - r_0)_1^2 - (r - r_0)_2^2} \times (\text{M.F.})^2 \text{ cm}^{-2} = \qquad \text{cm}^{-2}$$

$$\sigma^2 = -\frac{1}{2 \times \text{slope}}\text{cm}^2 = \qquad \text{cm}^2$$

	Top	Bottom
Slope:[10]	cm^{-2}	cm^{-2}
σ^2:	cm^2	cm^2
σ:	cm	cm

VII. Apparent Molecular Weights

$$M_{\text{app},0}^{(11)} = \frac{RT\rho_0}{\sigma^2\omega^2 r_0(d\rho/dr)_{\text{phys},0}}$$

Top: g/mole
Bottom: g/mole

rotor stretch and some typical values which may be used in lieu of experimentally determined values for a particular rotor.

[8]The density may be obtained in ways other than refractometry as discussed in the text. If refractometry is employed, the appropriate coefficients from Table VI are inserted here.

[9]In some cases, the calculations may end at this point if the molecular weight is not sought. In any event, the calculations must proceed to here if hydration parameters are to be reported. These pressured buoyant densities are from 0.006 to 0.009 density units higher than the corresponding values at atmospheric pressure depending on where in the column the band is formed, what angular velocity is employed and what salt is used.

[10]These slopes are not expected to be identical but, as expected from the discussion in section IV of the Ifft and Vinograd (1962) paper, to vary due to the non-constant gradient. The limiting slopes as $(r - r_0) \rightarrow 0$ should approach the correct value.

[11]As discussed at length in the theoretical section, molecular weights obtained from this relation can be expected to be too low by 10–30% if there is interaction of the macromolecule with either or both components of the solvent.

Computers

It will be obvious to many readers that much of what has been suggested in the preceding two sections can be done with far greater speed if computers are used. Because practically every protein research laboratory has immediate access to or can send programs and data to a nearby computing center, serious thought should be given to ways in which computations such as those outlined in Table VII can be speeded or additional data derived which is generally not sought due to a lengthy and laborious calculation.

The reader is referred to Trautman's (1966) highly readable and informative account of the contributions computers have made to ultracentrifugal analysis. He gives a general discussion of the capabilities of computers and programming languages and provides useful summaries of programs which have been developed for use with the moving boundary method with optical registration of boundary position, Archibald method for molecular weight from schlieren optics, sedimentation equilibrium method with Rayleigh optics and preparative centrifugation. It is clear from these examples that the relatively simple computation outlined in Table VII as well as more complicated treatments which include such effects as the non-constant density gradient can be readily expressed in Fortran IV and computation time reduced to virtually zero.

Bartulovich and Ward (1965) have gone one step beyond this to help alleviate the tedious procedure outlined above under "Measurement of the plates". They attached digital encoders to the two lead screws on a toolmaker's microscope and fed this output through data processors directly to a tape punch which provided the input tape for a computer. This may prove to be a highly satisfactory method of analyzing density gradient plates when a method can be found to interpolate through the areas of the schlieren image which are difficult to read.

TABLE VIII. ISOTHERMAL
COMPRESSIBILITY COEFFICIENTS*
OF SALT SOLUTIONS OF
DENSITY 1.3

Salt	$\kappa \times 10^6$, atm^{-1}	Salt	$\kappa \times 10^6$ atm^{-1}
CsCl	35.	CsBr	40.
KBr	33.	CsI	43.
RbBr	37.	Cs_2SO_4	34.

*The temperatures at which the compressibility coefficients were determined varied from 13.8° to 25°. Because the isothermal compressibility is an insensitive function of temperature, the values can be assumed to be valid at 25°.

B. Preparative Ultracentrifugation

The technique of sedimentation in a density gradient in preparative centrifuges has become of great importance in biochemical investigations. It will assume an even greater significance for equilibrium studies of proteins with the advent of the high-speed preparative instruments such as Spinco's Model L-2 and IEC's B-60 which are capable of speeds of 65,000 and 60,000 rpm respectively. The advantages are the much larger capacities which yield preparative amounts, the much lower costs and the wide spectrum of techniques which can be employed to analyze the fractions derived from the equilibrium distribution. The disadvantages in comparison to analytical ultracentrifugation are that no visual display is available during the experiment, no photometric record of the equilibrium distribution can be obtained, the preparative centrifuge tubes are not as sophisticated as the spectrum of sectorial centerpieces available and the experiment requires that the centrifugal field be totally relaxed to permit analysis.

1. Run Preparations and Conduct of the Run

Solutions, in considerably larger volumes, are prepared as above unless a preformed gradient is required. Procedures for the establishment of preformed gradients are given in numerous papers such as the classic paper by Martin and Ames (1961) and the chapter by Svensson (1960) in volume 1 of this series.

A swinging-bucket rotor is always required because of the necessity that the centrifugal force be directed radially outward along the axis of the tube during centrifugation and the gravitational field directed vertically downward along the axis of the tube while at rest. Examples of such rotors are the SW-39 for use in the Spinco Model L, the SW-39E for use in the Model E and the new SB-269 for use in the B-60 IEC centrifuge. The latter rotor is made from titanium and derating is never required.

The tubes are generally made from cellulose nitrate for ease in the analysis step. Polypropylene or stainless steel can be used if the solution is removed by aspiration. The tubes most commonly used are 5-ml although spacers are provided for smaller tubes. A layer of fluorocarbon on the bottom will assure a radial outer surface. If less than 5 ml of the salt solution are used, a few milliliters of mineral oil layered on top will prevent tube collapse at high speeds.

Somewhat longer times are required to reach equilibrium than in a comparable analytical experiment because of the longer column length. Run times of 48–60 hr are commonly employed to ensure equilibrium in CsCl solutions.

Maximum rated speeds are generally employed. However, Spinco suggests that for solutions more dense than 1.2 g/ml:

$$\text{Allowed maximum speed} = \sqrt{\frac{1.2 \text{ g/ml}}{\text{density of sample}}} \times \text{maximum rated speed.}$$

2. Analysis

A reasonable degree of manual dexterity is required to analyze a preparative density gradient run. The rotor must be removed from the chamber and the tube removed from the rotor without significant amounts of mixing occurring. Although some commercial sampling devices are available, frequently the techniques used will depend upon the particular system being studied and are devised by each investigator. For instance, if there is a contaminant at either the top or the bottom of the cell, this may preclude removal of the solution from the top or bottom respectively.

A wide variety of techniques have been described in the literature. These include:

(a) Insertion of a needle or pin in the bottom or side of the tube and removal of successive several-drop fractions.
(b) Aspirating small amounts at a time from the meniscus of the solution.
(c) Freezing the tube in a dry-ice bath followed by slicing horizontal sections.
(d) Injecting a solution of high density into the bottom of the tube and gradually displacing the solution out of the top of the tube.

Trautman (1964) provides useful schematic diagrams of these procedures.

Fractions obtained by one of the above methods can be analyzed for a wide range of properties: absorbance, generally at 260 and 280 mμ for nucleic acids and proteins respectively; radioactivity, P^{32} and C^{14}, are frequently employed for nucleic acids and proteins respectively; biochemical assay for enzymatic activity; refractive index; etc.

Ifft *et al.* (1961) presented data obtained by centrifuging CsCl solutions to equilibrium in a Spinco Model L centrifuge and analyzed the tube contents by bottom needle puncture. Their results, given in Fig. 18, are

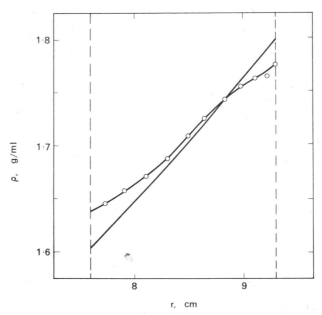

FIG. 18. Density distributions for a CsCl solution, $\rho_e = 1.700$, 2 ml, at equilibrium in a cylindrical tube in an SW-39 rotor in a Model L preparative ultracentrifuge at 39,000 rpm and 25°C. Solid line, theoretical distribution; O, experimental distribution. (Ifft *et al.*, 1961.)

similar to the density distributions found by Huang and Bonner (1965). These results show that severe departures from a smoothly increasing density function occur at the meniscus and cell bottom. This deviation was demonstrated to be due to back diffusion by an analytical experiment in which equilibrium was established at 39,460 rpm and then the field relaxed to 9945 rpm and the gradient curve examined as a function of time.

3. Information Obtainable

Frequently, the preparative experiments are designed solely to separate biological polymers and no properties of the polymer are determined from the experiment. Recently Huang and Bonner (1965) have accurately determined the density distribution by centrifuging one cell containing only the CsCl solvent and carefully analyzed its distribution by bottom needle puncture. The density of the polymer was obtained to three decimal places and an accurate hydration parameter could be obtained from this if $\bar{v}_{polymer}$ was known. Thus far, no attempts have been made in preparative experiments to quantitatively analyze the shape of the distribution for a molecular weight value.

IV. APPLICATIONS

A. Buoyant Behavior of Precipitated Proteins

Cox and Schumaker (1961) presented the first extensive report of an examination of proteins at sedimentation equilibrium in salt density gradients. They obtained absorption photographs of several proteins (see Fig. 19) and concluded that the buoyant density of dissolved proteins could not be obtained with satisfactory precision. Therefore they turned to precipitating solvents which were mixtures of CsCl and $(NH_4)_2SO_4$. Sharp bands were then obtained as indicated in Fig. 20. From such bands, buoyant densities were given to three decimal places. Their results are presented in Table IX. It is the most extensive published tabulation of buoyant densities of proteins to date. Unfortunately, it is not entirely clear how these densities were computed. They clearly correspond to values of ρ_0^0 as pressure effects were not considered in this early work.

The authors present convincing evidence for H-meromyosin that the ρ_0^0's of the precipitated and dissolved proteins are identical. The evidence for BSA is not so certain. Recently, Zilius and Ifft (unpublished) have demonstrated that for a wide spectrum of proteins, including conalbumin, whale myoglobin, cytochrome c and ovalbumin, the precipitated protein is more dense than the dissolved protein. The magnitude of the difference ranges from 0.004 to 0.031 g/ml. Figure 21, which displays the bands of precipitated and dissolved ovalbumin in CsCl, is typical.

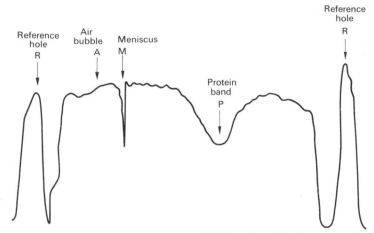

FIG. 19. Densitometer tracing of the equilibrium ultraviolet absorption photo-graph of dissolved bovine serum albumin in 2.8 molal CsCl at 59,780 rpm. (Cox and Schumaker, 1961.)

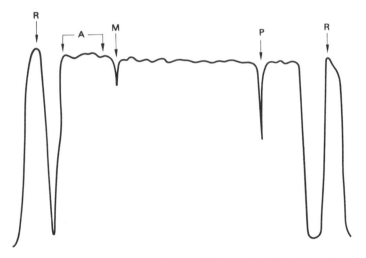

FIG. 20. Densitometer tracing of the equilibrium ultraviolet absorption photo-graph of precipitated lysozyme in 0.8 molal CsCl and 3.6 molal $(NH_4)_2SO_4$ at 59,780. The letters are explained in Fig. 19. (Cox and Schumaker, 1961.)

Cox and Schumaker (1961) give hydrations for each protein in the last column. These figures do not take into account the effects of pressure and anion binding, but they do provide values in the range obtained by other methods. The authors were the first to point out that values of hydration may be obtained from density gradient data by two independent measure-ments. Measurement of ρ_0 yields Γ' from Eq. (20). They also point out

TABLE IX. HYDRODYNAMIC DENSITIES AND APPARENT PREFERENTIAL
HYDRATION VALUES FOR PROTEIN PRECIPITATES IN CONCENTRATED
SALT SOLUTIONS

Protein	Salts	m	Initial density	ρ_0^0	$\bar{v}_p$	Γ'
Human hemoglobin	CsCl	2.8	1.306	1.295 ± 0.002	0.749	0.10
Chicken myosin	CsCl	2.4	1.300	1.280 ± 0.002	0.74	0.19
	$(NH_4)_2SO_4$	0.3				
L-Meromyosin	CsCl	2.2	1.304	1.280 ± 0.002	(0.74)	0.19
	$(NH_4)_2SO_4$	0.6				
H-Meromyosin	CsCl	2.1	1.301	1.285 ± 0.002	(0.74)	0.18
	$(NH_4)_2SO_4$	0.8				
BSA	CsCl	0.7	1.253	1.255 ± 0.002	0.734	0.31
	$(NH_4)_2SO_4$	3.6				
Lysozyme	CsCl	0.8	1.267	1.290 ± 0.002	0.722	0.24
	$(NH_4)_2SO_4$	3.6				
Lobster hemocyanin	CsCl	0.8	1.267	1.270 ± 0.002	0.740	0.22
	$(NH_4)_2SO_4$	3.6				

that the molecular weight obtained from Eq. (13) is the value for the
hydrated molecule. Thus, α, the number of moles of bound water per
mole of protein, can be computed from the relation:

$$M_h = M_p + \alpha M_w, \tag{35}$$

where M_h and M_p are the solvated and anhydrous molecular weights
respectively.

Cox and Schumaker also give values for the buoyant densities of soluble BSA in 3.6 M $(NH_4)_2SO_4$ solutions to which increasing amounts of
CsCl are added. The observed increase in buoyant density with increasing
amounts of CsCl is probably due to binding of Cl^- ions to additional sites
on the molecule accompanied by replacement of NH_4^+ in the secondary
layer by Cs^+. Competitive binding studies are needed to evaluate this
suggestion and that of Cox and Schumaker, that hydration decreases.

B. Determination of the Molecular Weight, Solvation and Ion-binding of a Protein

Ifft and Vinograd (1962 and 1966) have investigated the behavior of
one protein, bovine serum mercaptalbumin, in a variety of salt solutions.
The purpose of the papers was to investigate techniques requisite to
conducting and analyzing runs involving broad protein bands, to examine
quantitatively the effects of several approximations inherent in the derivation of the fundamental equations, to determine whether or not proteins
form Gaussian concentration bands and to determine whether the correct
molecular weight of a polymer of known molecular weight is obtained if
the effects of solvation and anion binding are included.

Ifft and Vinograd (1962) investigated a variety of methods of conducting these experiments and their results as to how to set up the run and measure the plates are given in Section III of this chapter. The Tabular Outline presented there gives the most satisfactory method of computing the standard deviation. Several other methods can be employed. If the baseline is subtracted from the solution gradient curve and the protein gradient curve replotted, the distance between band center and each maximum is equal to σ. Alternatively, if the area under the gradient curve is integrated and the data replotted to yield the concentration profile, the halfwidth at $1/\sqrt{e} = 0.607$ of the maximum height equals σ. Such a calculated distribution is plotted in Fig. 22. Both of these methods

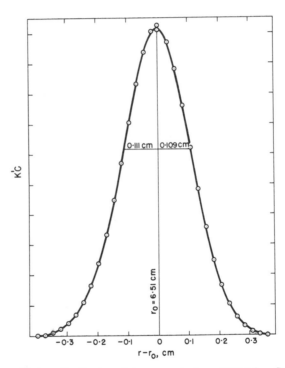

FIG. 22. Equilibrium distribution of the concentration of BMA in a CsCl gradient. Obtained by numerically integrating the areas under the protein gradient curve in Fig. 17. (Ifft and Vinograd, 1962.)

suffer because they rely on measurements at only one point in the distribution. A plot of $1/n \cdot dn/dr$ vs. $(r - r_0)$ can be employed but it is very sensitive to errors made at the edges of the band.

These workers concluded that using the techniques as described in Section III, the buoyant density could be determined to ± 0.001 g/ml, band center to ± 0.001 cm and the standard deviation to $\pm 2\%$.

Typical results of such experiments are displayed in Figs. 23 and 24. The former presents the computer-calculated density distribution, the density gradient distribution obtained from the ρ values and finally a tracing of the photographic plate. Figure 24 shows a plot of log k' $(n - n_0)$ vs. $(r - r_0)^2$ for the two lobes. The lines should be linear and superimposed. Because of several inconsistencies observed in Figs. 22, 23 and 24, a number of basic assumptions were re-examined.

The effect of a linear variation of the refractive index increment through the band and the effect of discarding terms of the order of $(r - r_0)^3$ in the original derivation were found to be negligible. The fact that the gradient

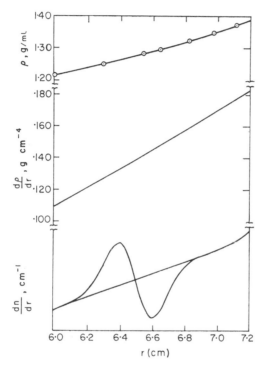

FIG. 23. Equilibrium density, density gradient and refractive index gradient for the experiment described in Fig. 17. (Ifft and Vinograd, 1962.)

is not constant as displayed in the middle diagram of Fig. 23 but varies almost linearly with distance with a slope of 0.060 g/cm⁵ was considered. The inclusion of the third-order term, 0.060 $\epsilon^2/2$ where $\epsilon = (r - r_0)$, in the Taylor expansion of ρ led to the relation:

$$\ln c/c_0 = -\frac{1}{2\sigma^2}\epsilon^2(1 + 0.25\,\epsilon).\qquad(36)$$

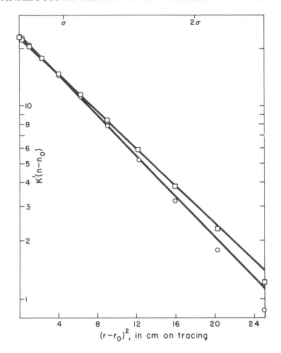

FIG. 24. Logarithmic plot of the concentration curve derived in Fig. 17. ○, outer half of distribution; □, inner half of distribution. (Ifft and Vinograd, 1962.)

The equation predicts that data on the logarithmic plot from the top half of the distribution should curve upwards and vice versa for the data from the lower half of the distribution. This artifact also explains the skewed distribution observed in Fig. 22. The last perturbation considered was the effect of making a small error, δ, in the numerical integration at the outer edge of the band assuming that the integration begins there. The effect is not small and is noticed as a uniform elevation or depression, depending upon the sign of δ, in the logarithmic plot. The authors concluded that accurate values of σ can only be obtained by beginning the numerical integration at band center and working outwards and by taking the limiting logarithmic slopes as $(r - r_0) \rightarrow 0$.

Apparent specific volumes were determined in the CsCl solution of buoyant composition for use in the solvation computation. Surprisingly, a value of 0.736 was obtained which is nearly identical to the 0.734 value which was found in water. This value, the value of the pressured buoyant density, and an assumed partial specific volume of 1.00 for the bound water led to a Γ' of 0.200 g H_2O/g protein.

The physical density gradient was used to compute a solvated molecular weight of 80,100. This does not correspond well with the value of 84,000 obtained by Eq. (35). This result plus the well-known behavior of BMA in binding anions led to an extension of the investigation to several other salt systems.

Ifft and Vinograd (1966) determined the buoyant density and standard deviations of the bands formed by BMA in six salt solutions. The salts investigated included the halide series, CsI, CsBr and CsCl, the alkali metal series, KBr, RbBr and CsBr, and one independent salt, Cs_2SO_4. Buoyant experiments similar to the ones described above for CsCl were conducted for KBr and RbBr because β (ρ) values were available. Density gradients in the other salt solutions were obtained by simultaneously running two cells with two solutions of the same salt of densities bracketing the buoyant density. Photos similar to Fig. 16 were obtained. The properties of the gradient obtained by simply dividing the difference in the initial solution densities by the distance between the bands were examined by the authors. This gradient was termed the buoyancy* density gradient. Its properties along with those of the three other density gradients of interest are tabulated in Table X. Values of ρ_0^0 were computed from either

TABLE X. FOUR DENSITY GRADIENTS OF SIGNIFICANCE
IN DENSITY GRADIENT EXPERIMENTS

Name	Definition
Composition	$(d\rho/dr)_{comp} = \dfrac{\omega^2 r}{\beta^0}$
Physical	$(d\rho/dr)_{phys} = \left(\dfrac{1}{\beta^0} + \kappa\rho^{02}\right)\omega^2 r$
Buoyancy	$(d\rho/dr)_{buoy} = \left[\dfrac{1}{\beta^0} + \dfrac{(\kappa - \kappa_s)}{(1-\alpha)}\rho^{02}\right]\omega^2 r$
Effective	$(d\rho/dr)_{eff} = \left[\dfrac{(1-\alpha)}{\beta^0} + (\kappa - \kappa_s)\rho^{02}\right]\omega^2 r$

the composition gradient or the buoyancy gradient, the radial banding position and either known or interpolated isoconcentration positions. These buoyant densities are given in order of increasing density in the first column of Table XI. These densities were corrected to ρ_0 values with the isothermal compressibility coefficients given in Table VIII.

Because BMA is known to bind large numbers of anions, the anion binding in the salt solutions of buoyant composition were measured by the method of Scatchard and Black (1949). The results are given in column 4 of Table XI.

*The authors have revised this term slightly. Instead of the adjective buoyant, the term buoyancy will be used hereafter.

TABLE XI. RESULTS OF BUOYANT EXPERIMENTS WITH BOVINE SERUM
MERCAPTALBUMIN IN SEVERAL SALT SOLUTIONS
$\omega = 56,100$ rpm; $25.0°C$

Salt	ρ_0^0	ρ_0	Number of anions bound	Γ_*'	$\rho_{0,*}$	σ(cm)	$(d\rho/dr)_{eff}$	$M_{s,0}$	M_a
Cs_2SO_4	1.237	1.241	25	0.78	1.17	0.071	0.195	137,000	72,000
CsCl	1.278	1.282	53	0.51	1.21	0.106	0.120	102,000	63,000
KBr	1.295	1.302	67	0.37	1.24	0.147	0.064	103,000	69,000
RbBr	1.302	1.310	66	0.46	1.22	0.104	0.115	112,000	70,000
CsBr	1.306	1.315	70	0.58	1.20	0.084	0.165	112,000	68,000
CsI	1.331	1.347	78	0.61	1.20	0.071	0.211	133,000	70,000

Because the protein is banded in a buffered solution of pH 5.5 which is assumed to be the isoelectric point of the protein in this concentrated salt solution, the binding of x number of Cl^- ions must result in a concomitant binding of x number of Cs^+ ions in a secondary layer to provide charge neutrality. This will increase the density of the buoyant species and hence must be accompanied by an increased solvation. A new hydration parameter, Γ_*' was therefore formulated as the *net hydration of the protein-salt complex* and was computed from Eq. (30). These hydrations are tabulated in column 5 of Table XI. They are considerably higher than the *net hydration of the salt-free protein*, Γ', values obtained from Eq. (20).

The *derived buoyant density*, $\rho_{0,*}$, was defined as:

$$\rho_{0,*} = \frac{1 + \Gamma_*'}{\bar{v}_3 + \Gamma_*' \bar{v}_1}. \qquad (37)$$

This is the density the protein would exhibit if no salt were bound and the solvation were not affected by salt binding.

If the theory of Hearst and Vinograd (1961) is to be successfully applied to proteins, it must be demonstrated that the $\bar{v}_{s,0}$ depends not only on pressure but water activity as well. In the light of the above discussion, Γ' and ρ_0 values would not be expected to vary monotonically with the activity of water in the solution. Values of Γ_*' and $\rho_{0,*}$ should exhibit this dependence and Fig. 25 demonstrates that this is indeed the case. Effective density gradients were computed for each salt solution and from these, solvated albumin-salt complex molecular weights and anhydrous molecular weights were calculated. These data are presented in the last three columns of Table XI. The corresponding z-average molecular weight obtained by two-component sedimentation equilibrium was 70,500. It is evident that while density gradient centrifugation is not the most straight-

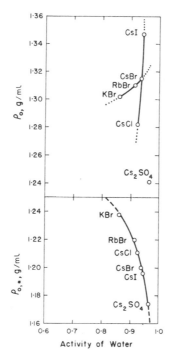

F IG. 25. The measured and derived buoyant densities of BMA in several salt solutions as a function of water activity. (Ifft and Vinograd, 1966.)

forward method for determining molecular weights, it nevertheless is capable of doing so. This investigation provides the first demonstration of this fact.

C. Resolution of Isotopically Labeled Proteins

Differentiation of pre-existing and newly-synthesized enzymes is of considerable biochemical interest. Hu et al. (1962) have demonstrated that preparative density gradient centrifugation can be used successfully in such experiments.

They grew E. coli ML 308 in a normal medium and then in media containing the stable isotopes D^2, N^{15} and C^{13} and combinations thereof. The enzyme β-galactosidase was obtained in partially purified form from the ruptured cells. An 0.1-ml sample of the enzyme preparation was layered on a pre-cooled, exponential gradient of RbCl. The tubes were placed in an SW-39 rotor and spun at 30,000 rpm for 80 hr. Following centrifugation, the distribution of enzyme through the tube was analyzed by bottom needle puncture and each fraction assayed for enzymatic activity.

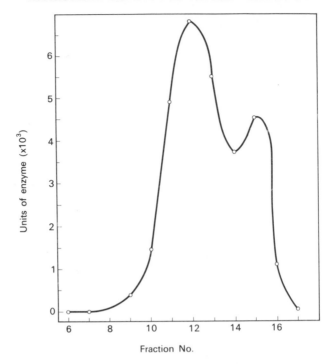

FIG. 26. Equilibrium distribution of N^{15} labeled β-galactosidase and unlabeled enzyme in a RbCl gradient in a preparative centrifuge after 75 hr at 500 rps. Density increases with increasing fraction number. (Hu *et al.*, 1962.)

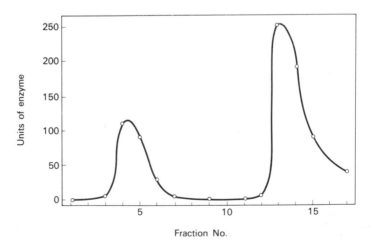

FIG. 27. Equilibrium distribution of D^2, N^{15}-doubly labeled β-galactosidase and unlabeled enzyme in a RbCl gradient in a preparative centrifuge at 500 rps. Density increases with increasing fraction number. (Hu *et al.*, 1962.)

Figures 26 and 27 demonstrate the resolution which was achieved. The N^{15} labeled enzyme did not differ in density sufficiently from the unlabeled enzyme to achieve a satisfactory separation. However, the doubly labeled D^2, N^{15} enzyme was completely resolved from the unlabeled protein. The authors point out that there are large enough differences in density of natural proteins such as serum albumin and fibrinogen that they could be similarly separated.

Another interesting observation made was that separations as large as 90% could be effected in centrifuge tubes containing very shallow gradients and proteins of slightly differing densities. The experiments would not yield discrete bands of protein but would generate modified two-component sedimentation equilibrium distributions.

D. Lipoproteins

Adams and Schumaker (1964) have reported a study of human plasma lipoproteins in the Model E ultracentrifuge. The mixture of lipoproteins was banded in a sucrose–sodium bromide solution. Equilibrium was achieved in 20–24 hr. As would be expected from the above discussion, the molecules banded at densities considerably less than the density of 1.3 exhibited by most proteins. The lipoproteins were found to band at densities ranging from 1.019 to 1.063.

One highly unsymmetrical gradient curve was analyzed by reflecting the two extremes of the curve and subtracting from the observed curve to yield what may have been the original distribution. Two symmetrical of 1.3 exhibited by most proteins. The lioporoteins were found to band low-density class of lipoproteins consists of only two components, having densities of 1.026 and 1.030 g/ml.

E. Discovery of Histone-bound RNA

Equilibrium experiments in density gradients of necessity offer an environment which proteins are not frequently exposed to during physical-chemical analyses. This variable is a very high ionic strength, $\mu = 2$ or higher. Huang and Bonner (1965) have utilized this property to excellent advantage to provide the exciting discovery of histone-bound RNA.

The usual procedure to purify histone from chromatin has employed acid. Pure histones were obtained because the histone–RNA bond is cleaved under these conditions. Huang and Bonner used a CsCl density gradient to remove the histones from the native nucleohistone, which has a much higher density.

Pea seedlings were grown from seeds which had been soaked in a P^{32} medium and large quantities of apical buds harvested. Chromatin was

separated from the starch by pelleting. Soluble nucleohistone was prepared by sucrose centrifugation, dialysis and shearing. The nucleohistone
solution was dissolved in 2.09 M CsCl at pH 8.0 and centrifuged at
39,000 rpm in the Spinco Model L centrifuge in an SW-39 rotor for 48 hr.
Analysis was by bottom needle puncture using the Buchler dripping
device. The needle was inserted far enough so that none of the pelleted
DNA was removed. The protein content of each fraction was analyzed
by determining $O.D._{230m\mu}$ and the nucleic acid content by counting P^{32}
activity. The density distribution was determined in each run by running
a blank CsCl tube and measuring the refractive index of each fraction.
The results of such an experiment are given in Fig. 28. It is apparent
that a major component of this preparation is a nucleic acid bound to
a protein, the complex having a density of 1.286. A minor component of
$\rho_0^0 = 1.256$ can also be identified.

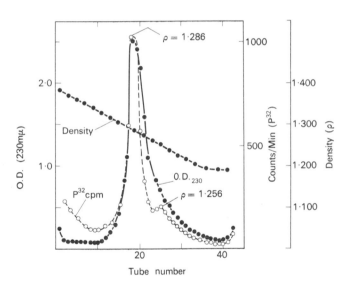

FIG. 28. Equilibrium distribution of histone-RNA from pea buds dissolved in
2.09 M CsCl, pH 8.0, tris buffer ($\mu = 0.01$) after 48 hr at 39,000 rpm and 20°C.
(Huang and Bonner, 1965.)

This material was subjected to further chemical, electrophoretic and
chromatographic analyses which demonstrated that the protein is
definitely a histone and the nucleic acid is RNA. It was further shown
that the RNA contains thirty-eight nucleotides. This information plus
the mass ratios of RNA/histone and the known molecular weight of the
histone demonstrate that there must be six to twelve histone molecules
associated in a larger unit.

CsCl-centrifugation was also employed to demonstrate that the histone isolated above was indeed different from the protein prepared by the acid treatment. Native histone separated in 0.2 N H_2SO_4 displayed two bands of densities 1.240 and 1.218, neither of which correspond with the density of the RNA-histone complex.

The authors demonstrated that the two proteins exhibiting the bands in Fig. 28 could be resolved by treatment with 0.4 M $NaClO_4$. Rebanding of the RNA-protein after such treatment revealed only one band of $\rho_0^0 =$ 1.286.

Interestingly, the authors satisfactorily computed the expected buoyant densities of the two classes of nucleo-histones from their RNA contents, the known density of RNA (1.900) and the mean density of the two pure histones (1.229) using the relation:

$$\frac{1}{\rho} = \frac{W_1}{\rho_1} + \frac{W_2}{\rho_2},$$

where W_1 and W_2 are the weight fractions of the two components of density ρ_1 and ρ_2.

F. Phase Transitions of Collagen

Fessler and Hodge (1962) demonstrated another unique application of the density gradient technique to the study of protein structure. They pointed out that the method is ideally suited for the study of phase transitions provided that the densities of the two phases differ and that the transition temperature lies within the range of the instrument (0–40°C unless the centrifuge is equipped with the high temperature unit).

Neutral salt-soluble collagen was banded in a CsBr gradient at pH 7.3 at 6°C. A normal schlieren photograph similar to Fig. 17 was obtained. An approximate molecular weight was calculated by the method of Ifft and Vinograd (1962, 1966). The value computed corresponded closely to that of monomeric tropocollagen.

At 12°C, a hypersharp band was formed near the minimum of the protein gradient curve at a density 0.012 g/ml greater than the soluble collagen. This photograph was quite similar to that of soluble and precipitated conalbumin in Fig. 21. It indicates the aggregation of collagen into fibrils. On further warming to 22°C, the same two bands were noted. The amount of soluble collagen had decreased and the concentration of fibrils had increased.

At 31°C, no soluble collagen remained. The sharp fibril band remained and a new broad band was formed. The new band had the same density as the parent gelatin and a band width consistent with a separation of the three-stranded helix into monomeric units. Upon cooling this material to 6°C, a band was formed at the same density as soluble collagen but it remained in the aggregated state.

These results can be explained as a loss of a hydration shell as the soluble collagen molecules aggregate into fibres or as an increase in salt binding.

G. Buoyant Titration of Bovine Serum Mercaptalbumin

Williams and Ifft (unpublished) have employed density gradient centrifugation in CsCl solutions in an attempt to locate the anion binding sites of BMA. This unique protein binds large numbers of anions. Numerous studies have been made to determine the number of anions bound under various conditions and the binding constants of each class of sites. However, there has been little definitive work to ascertain where on the molecule such large numbers (up to 60) of anions are bound. These authors decided that a selective chemical modification of the protein followed by remeasurement of the anion binding could yield the desired result. The modification chosen was titration of the ionizable residues and the method of measurement selected was density gradient centrifugation.

Figure 29 gives the buoyant density of BMA in CsCl at 25°C as a function of pH. The size of the circles denotes the uncertainty in ρ_0 in each of the measurements. The increase in ρ_0 as the pH is raised can be

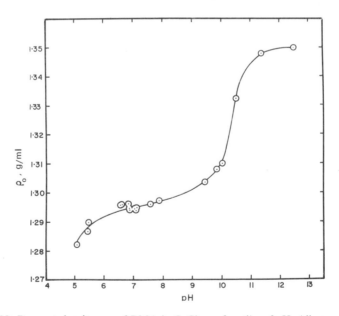

Fig. 29. Buoyant density, ρ_0, of BMA in CsCl as a function of pH. All runs performed in 4°, single sector, Kel–F centerpieces at 56,100 rpm and 25°C. (Williams and Ifft, in preparation.)

qualitatively assigned to a loss of water or an increase in anion or cation binding.

In order to quantitatively interpret the data, the authors expressed the buoyant density as an extension of Eq. (20):

$$\rho_0 = \frac{1 + z_{Cs^+} + z_{Cl^-} + \Gamma'_*}{\bar{v}_3 + z_{Cs^+}\bar{v}_{Cs^+} + z_{Cl^-}\bar{v}_{Cl^-} + \Gamma'\bar{v}_1} \qquad (38)$$

where z_i and $\bar{v}_i$ are the grams of i bound per grain of protein and partial specific volume of i respectively. This form of the equation is required rather than Eq. (30) because as the pH is changed from the isoelectric condition, the number of bound cesium and chloride ions can no longer be equal. The electroneutrality condition was expressed as:

$$\text{Net negative charge on BMA} = \nu_{Cl^-} - \nu_{Cs^+}, \qquad (39)$$

where ν_i is the number of ions bound/molecule BMA.

These two equations contain three variables. One of the variables, Γ'_*, was determined at the isoelectric pH and at pH 12.5. It was computed at pI with Eq. (30) at 0.48 g H_2O/g protein. Binding data indicate that there are no chloride ions bound at pH 12.5. Therefore by electroneutrality and knowing the number of ionizable groups, it was concluded that there were 106 Cs^+ bound per molecule and a Γ'_* of 0.45 was determined. Because the hydration changes so slightly over such a large pH range, which includes a large structural transformation (Leonard et al., 1963), the value of Γ'_* was taken as a slowly varying, linear function from pH 5.5 to pH 12.5.

Tanford's (1955) values for the intrinsic pK's of the ionizable groups of BMA were used to compute the net charge on BMA at pH intervals 0.25 pH units apart over the pH range of interest. Partial specific volumes recently computed by Ifft and Williams (1967) for cesium and chloride ions in solution were employed. The partial specific volume of BMA was taken as 0.736 as measured by Ifft and Vinograd (1962) and the partial specific volume of water was assumed to be 1.00. Then Eqs. (38) and (39) were combined to yield the ν_{Cs^+} and ν_{Cl^-}. The results of this work are given in Fig. 30.

Scatchard (1964) has determined that there are twenty-two Cl^- binding sites in the third most active class of sites of serum albumin. Comparison of the large inflection point in Fig. 30 of the ν_{Cl^-} curve at pH 9.6 with Tanford's value of 9.8 for the intrinsic dissociation constant of the lysine residues of serum albumin indicates that twenty-two of the fifty-seven ϵ-NH_3^+ groups of the lysine residues constitute this third class of binding sites. The sharp rise in ν_{Cs^+} at pH 10.5 is attributed to the binding of a Cs^+ to each of the eighteen tyrosine residues of BMA.

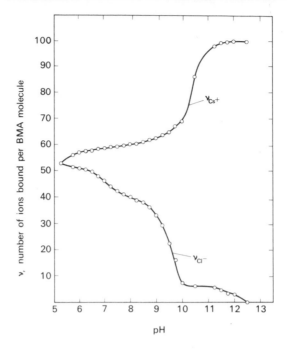

FIG. 30. Number of cesium and chloride ions bound per molecule of BMA as a function of pH. (Williams and Ifft, in preparation.)

H. Behavior of Poly-L-Glutamic Acid in Density Gradients

Vinograd and Hearst (1962) reported the first buoyant density titration (unpublished data of J. Vinograd and J. Morris). The cesium salt of PGA exhibits a constant density of 1.70 g/ml above pH 7. The density decreases from pH 7.0 to pH 5.1 when cesium ions are removed as the carboxyls are protonated. The polypeptide precipitates at pH 5.1 in these concentrated salt solutions. The buoyant density of the precipitated material decreases to 1.50 at pH 1.0.

ACKNOWLEDGEMENTS

The author is very pleased to acknowledge the generous assistance of Professors Jerome Vinograd and John Hearst in learning about density gradient centrifugation and the several helpful discussions during the preparation of this manuscript.

APPENDIXES

Appendix A

Ultracentrifuge Protocol   Log number ....
Density Gradient Sedimentation   Operator   ....
                                                           Date   ....

Protein ....

Solution placement

Soln. no.	Cell no.	Sector	Window combination	Ml	Protein conc.	Comments
1						
2						
3						
4						

Solution properties

	Initial			Final		
	n_D^{25}	ρ^{25}	pH	n_D^{25}	ρ^{25}	pH
Solution # 1						
Solution # 2						
Solution # 3						
Solution # 4						

Centrifuge data

Angular velocity   ....   Temperature   ....
Rotor type   ....   Time   ....
Rotor serial no.   ....   Odometer   ....
Control voltage   ....   Film   ....
Evapotrol setting ....   Developer.   ....

Day	Date	Time	Plate-frame	Exp.-sec	Interval	ϕ	RTIC balance

Appendix B

THE RELATION BETWEEN REVOLUTIONS PER MINUTE AND ANGULAR VELOCITY SQUARED

(rpm values correspond to gearbox settings of Spinco Model E analytical ultracentrifuge)

RPM (rev/min)	ω^2 (radians/sec)^2	RPM (rev/min)	ω^2 (radians/sec)^2
1967	$4 \cdot 2429 \times 10^4$	12590	$1 \cdot 7382 \times 10^6$
2095	$4 \cdot 8131 \times 10^4$	13410	$1 \cdot 9720 \times 10^6$
2233	$5 \cdot 4681 \times 10^4$	14290	$2 \cdot 2393 \times 10^6$
2378	$6 \cdot 2013 \times 10^4$	15220	$2 \cdot 5403 \times 10^6$
2531	$7 \cdot 0249 \times 10^4$	16200	$2 \cdot 8780 \times 10^6$
2695	$7 \cdot 9648 \times 10^4$	17250	$3 \cdot 2631 \times 10^6$
2809	$8 \cdot 6529 \times 10^4$	17980	$3 \cdot 5452 \times 10^6$
2994	$9 \cdot 8302 \times 10^4$	19160	$4 \cdot 0258 \times 10^6$
3189	$1 \cdot 1152 \times 10^5$	20410	$4 \cdot 5682 \times 10^6$
3397	$1 \cdot 2655 \times 10^5$	21740	$5 \cdot 1829 \times 10^6$
3617	$1 \cdot 4346 \times 10^5$	23150	$5 \cdot 8770 \times 10^6$
3848	$1 \cdot 6238 \times 10^5$	24630	$6 \cdot 6525 \times 10^6$
4059	$1 \cdot 8067 \times 10^5$	25980	$7 \cdot 4018 \times 10^6$
4327	$2 \cdot 0532 \times 10^5$	27690	$8 \cdot 4082 \times 10^6$
4609	$2 \cdot 3295 \times 10^5$	29500	$9 \cdot 5434 \times 10^6$
4908	$2 \cdot 6416 \times 10^5$	31410	$1 \cdot 0819 \times 10^7$
5227	$2 \cdot 9961 \times 10^5$	33450	$1 \cdot 2270 \times 10^7$
5563	$3 \cdot 3937 \times 10^5$	35600	$1 \cdot 3898 \times 10^7$
5784	$3 \cdot 6687 \times 10^5$	37020	$1 \cdot 5029 \times 10^7$
6166	$4 \cdot 1693 \times 10^5$	39460	$1 \cdot 7075 \times 10^7$
6569	$4 \cdot 7321 \times 10^5$	42040	$1 \cdot 9381 \times 10^7$
6995	$5 \cdot 3658 \times 10^5$	44770	$2 \cdot 1980 \times 10^7$
7447	$6 \cdot 0816 \times 10^5$	47660	$2 \cdot 4910 \times 10^7$
7928	$6 \cdot 8926 \times 10^5$	50740	$2 \cdot 8233 \times 10^7$
8225	$7 \cdot 4187 \times 10^5$	52640	$3 \cdot 0387 \times 10^7$
8766	$8 \cdot 4268 \times 10^5$	56100	$3 \cdot 4513 \times 10^7$
9341	$9 \cdot 5685 \times 10^5$	59780	$3 \cdot 9189 \times 10^7$
9945	$1 \cdot 0846 \times 10^6$	63650	$4 \cdot 4428 \times 10^7$
10589	$1 \cdot 2296 \times 10^6$	67770	$5 \cdot 0365 \times 10^7$
11573	$1 \cdot 4688 \times 10^6$	74070	$6 \cdot 0165 \times 10^7$

ADAMS G. H. and SCHUMAKER V. N. (1964) Analytical equilibrium density gradient ultracentrifugation of human plasma lipoproteins. *Nature* **202**, 490–491.

ARCHIBALD W. J. (1947) A demonstration of some new methods of determining molecular weights from the data of the ultra centrifuge. *J. Phys. Coll. Chem.* **51**, 1204–1214.

BALDWIN R. L. (1959) Equilibrium sedimentation in a density gradient of materials having a continuous distribution of effective densities. *Proc. Natl. Acad. Sci. U.S.* **45**, 939–944.

BALDWIN R. L. and SHOOTER E. M. (1963) Measurement of density heterogeneity by sedimentation in preformed gradients. *Ultracentrifugal Analysis in Theory and Experiment*, edited by WILLIAMS J. W., pp. 143–168, Academic Press, New York.

BARTULOVICH J. J. and WARD W. H. (1965) Recording, plate-measuring system for the ultracentrifuge. *Anal. Biochem.* **11**, 42–47.

BECKMAN INSTRUMENTS, INC. (1960) An introduction to density gradient centrifugation. *Technical Review No. 1*. Spinco Division, Palo Alto.

BECKMAN INSTRUMENTS, INC. Model E analytical ultracentrifuge, *Instruction Manual E-IM-3*. Spinco Division, Palo Alto.

CLAESSON S. and MORING-CLAESSON I. (1961) Ultracentrifugation. *A Laboratory Manual of Analytical Methods of Protein Chemistry*, **3**, edited by ALEXANDER P. and BLOCK R. J., pp. 119–191, Pergamon Press, New York.

COX D. J. and SCHUMAKER V. N. (1961) The preferential hydration of proteins in concentrated salt solutions. II. Sedimentation equilibrium of proteins in salt density gradients. *J. Amer. Chem. Soc.* **83**, 2439–2445.

DAYANTIS J. and BENOIT H. (1964a) Influence de la concentration et de la polydispersité sur l'équilibre de sédimentation dans un gradient de densité. *J. Chim. Phys.* 773–780.

DAYANTIS J. and BENOIT H. (1964b) Influence de la concentration et de la polydispersité sur l'équilibre de sédimentation en gradient de densité. *Ibid.* 781–789.

DEDUVE C., BERTHET J. and BEAUFAY H. (1959) Gradient centrifugation of cell particles theory and applications. *Progress in Biophysics and Biophysical Chemistry*, **9**, edited by BUTLER J. A. V. and KATZ B., pp. 325–369, Pergamon Press, New York.

ENDE H. A. (1964) New cell designs for density gradient centrifugation. *Macromol. Chem.* **78**, 140–145.

ENDE H. A. (1965) Molecular weight averages of polymers as determined by density gradient centrifugation I. Application to various polystyrenes. *Ibid.* **88**, 159–178.

FESSLER J. H. and HODGE A. J. (1962) Ultracentrifugal observation of phase transitions in density gradients. *J. Mol. Biol.* **5**, 446–449.

FUJITA H. (1962) *Mathematical Theory of Sedimentation Analysis*, pp. 258–262, Academic Press, New York.

GIBSON R. E. (1934) The influence of concentration on the compressions of aqueous solutions of certain sulfates and a note on the representation of the compressions of aqueous solutions as a function of pressure. *J. Am. Chem. Soc.* **56**, 4–14.

GIBSON R. E. (1935) The influence of the concentration and nature of the solute on the compressions of certain aqueous solutions. *Ibid.* **57**, 284–293.

GOLDBERG R. J. (1953) Sedimentation in the ultracentrifuge. *J. Phys. Chem.* **57**, 194–202.

GOOD N. E., WINGET G. D., WINTER W., CONNOLLY T. N., IZAWA S. and SINGH R. M. M. (1966) Hydrogen ion buffers for biological research. *Biochemistry* **5**, 467–477.

GUCKER F. T., CHERNICK C. L. and ROY-CHOWDHURY P. (1966) A frequency-modulated ultrasonic interferometer: adiabatic compressibility of aqueous solutions of NaCl and KCl at 25°C. *Proc. Natl. Acad. Sci. U.S.*, **55**, 12–19.

HAUROWITZ F. (1963a) *The Chemistry and Function of Proteins*, p. 119, Academic Press, New York.

HAUROWITZ F. (1963b) *Ibid.*, pp. 111–121, 234–245.

HEARST J. E. (1961) Doctoral dissertation, p. 122. California Institute of Technology, Pasadena, California.

HEARST J. E. and VINOGRAD J. (1961a) A three-component theory of sedimentation equilibrium in a density gradient. *Proc. Natl. Acad. Sci. U.S.* **47**, 999–1004.

HEARST J. E. and VINOGRAD J. (1961b) The net hydration of T-4 bacteriophage deoxyribonucleic acid and the effect of hydration on buoyant behaviour in a density gradient at equilibrium in the ultracentrifuge. *Ibid.* 1005–1014.

HEARST J. E., IFFT J. B. and VINOGRAD J. (1961) The effects of pressure on the buoyant behaviour of deoxyribonucleic acid and tobacco mosaic virus in a density gradient at equilibrium in the ultracentrifuge. *Ibid.* 1015–1025.

HU A. S. L., BOCK R. M. and HALVORSON H. O. (1962) Separation of labeled from un-labeled proteins by equilibrium density gradient sedimentation. *Anal. Biochem.* 4, 489–504.

HUANG R. C. and BONNER J. (1965) Histone-bound RNA, a component of native nucleo-histone. *Proc. Natl. Acad. Sci. U.S.*, 54, 960–967.

IFFT J. B. and VINOGRAD J. (1962) The buoyant behavior of bovine serum mercaptalbumin in salt solutions at equilibrium in the ultracentrifuge. I. The protein concentration dis-tribution by schlieren optics and the net hydration in CsCl solutions. *J. Phys. Chem.* 66, 1990–1998.

IFFT J. B. and VINOGRAD J. (1966) The buoyant behavior of bovine serum mercaptalbumin in salt solutions at equilibrium in the ultracentrifuge. II. Net hydration, ion binding, and solvated molecular weight in various salt solutions. *Ibid.* 70, 2814–2822.

IFFT J. B., VOET D. H. and VINOGRAD J. (1961) The determination of density distributions and density gradients in binary solutions at equilibrium in the ultracentrifuge. *Ibid.* 65, 1138–1145.

IFFT J. B. and WILLIAMS A. (1967) The partial molar volumes of cesium and chloride ions in solution as a function of concentration. *Biochim. Biophys. Acta* (in press).

International Critical Tables, 3 (1928), edited by WASHBURN E. W., McGraw-Hill, New York.

KENCHINGTON A. W. (1960) Analytical information from titration curves. *A Laboratory Manual of Analytical Methods of Protein Chemistry*, 2, edited by ALEXANDER P. and BLOCK R. J., p. 361, Pergamon Press, New York.

LEONARD W. J. JR., VIJAI K. K. and FOSTER J. F. (1963) A structural transformation in bovine and human plasma albumins in alkaline solutions as revealed by rotatory disper-sion studies. *J. Biol. Chem.* 238, 1984–1988.

LUDLUM D. B. and WARNER R. C. (1965) Equilibrium centrifugation in cesium sulfate solutions. *J. Biol. Chem.* 240, 2961–2965.

LYONS P. A. and RILEY J. F. (1954) Diffusion coefficients for aqueous solutions of calcium chloride and cesium chloride at 25°. *J. Am. Chem. Soc.* 76, 5216–5220.

MAGDOFF B. S. (1960) Electrophoresis of proteins in liquid media. *A Laboratory Manual of Analytical Methods of Protein Chemistry*, 2, edited by ALEXANDER P. and BLOCK R. J., p. 202, Pergamon Press, New York.

MAHLER H. R. and CORDES E. H. (1966) *Biological Chemistry*, pp. 145–149 and 392–393, Harper & Row, New York.

MARTIN R. G. and AMES B. N. (1961) A method for determining the sedimentation behavior of enzymes: application to protein mixtures. *J. Biol. Chem.* 236, 1372–1379.

MERCK AND CO., INC. (1960) The Merck index of chemicals and drugs, pp. 1561–1566, Rahway, New Jersey.

MESELSON M., STAHL F. W., and VINOGRAD J. (1957) Equilibrium sedimentation of macro-molecules in density gradients. *Proc. Natl. Acad. Sci. U.S.* 43, 581–588.

MESELSON M. and NAZARIAN G. M. (1963) The transient state in density-gradient centri-fugation. *Ultracentrifugal Analysis in Theory and Experiment*, edited by WILLIAMS J. W., pp. 131–142, Academic Press, New York.

NATIONAL CANCER INSTITUTE MONOGRAPH, NO. 21 (1966) The development of zonal centrifuges, edited by ANDERSON N. G., U.S. Dept. Health, Education and Welfare, Bethesda.

PERLMAN G. E. and LONGSWORTH L. G. (1948) The specific refractive increment of some purified proteins. *J. Am. Chem. Soc.* 70, 2719–2724.

POHL F. (1906) Dissertation, Reinishe Friedrich-Wilhelms Universität, Bonn, Germany.

POLSON A. (1960) Multi-membrane electrodecantation. *A Laboratory Manual of Analytical Methods of Protein Chemistry*, 1, edited by ALEXANDER P. and BLOCK R. J., pp. 178–180, Pergamon Press, New York.

SCATCHARD G. and BLACK E. S. (1949) The effect of salts on the isoionic and isoelectric points of proteins. *J. Phys. Colloid Chem.* 53, 88–99.

SCATCHARD G. and YAP W. T. (1964) The physical chemistry of protein solutions. XII. The effects of temperature and hydroxide ion on the binding of small anions to human serum albumin. *J. Am. Chem. Soc.* **86**, 3434–3438.

SCHACHMAN H. K. (1959) *Ultracentrifugation in Biochemistry*, pp. 19–21, Academic Press, New York.

SOBOTKA H. and TRURNIT H. J. (1961) Uniomolecular layers in protein analysis. *A Laboratory Manual of Analytical Methods of Protein Chemistry*, **3**, edited by ALEXANDER P. and BLOCK R. J., pp. 211–243, Pergamon Press, New York.

SUEOKA N. (1959) A statistical analysis of deoxyribonucleic acid distribution in density gradient centrifugation. *Proc. Natl. Acad. Sci. U.S.* **45**, 1480–1490.

SVEDBERG T. and FÅHRAEUS R. (1926) A new method for the determination of the molecular weight of the proteins. *J. Am. Chem. Soc.* **48**, 430–438.

SVEDBERG T. and PEDERSEN K. O. (1940) *The Ultracentrifuge*, pp. 1–478, Oxford at the Clarendon Press. Reprinted by Johnson Reprint Corporation, New York.

SVENSSON H. (1960) Zonal density gradient electrophoresis. *A Laboratory Manual of Analytical Methods of Protein Chemistry*, **1**, edited by ALEXANDER P. and BLOCK R. J., pp. 211–220, Pergamon Press, New York.

SZYBALSKI W. (1967) Nucleic acids. Methods of enzymology, edited by GROSSMAN L. and MOLDAVE K., Academic Press, New York (in press).

TANFORD C. (1961) *Physical Chemistry of Macromolecules*, p. 358, Wiley, New York.

TANFORD C., SWANSON S. A. and SHORE W. S. (1955) Hydrogen ion equilibria of bovine serum albumin. *J. Am. Chem. Soc.* **77**, 6414–6421.

TODD A. (1960) Optical rotation. *A Laboratory Manual of Analytical Methods of Protein Chemistry*, **2**, edited by ALEXANDER P. and BLOCK R. J., pp. 245–283, Pergamon Press, New York.

TRAUTMAN R. (1960) Determination of density gradients in isodensity equilibrium ultracentrifugation. *Arch. Biochem. and Biophys.* **87**, 289–292.

TRAUTMAN R. (1964) Ultracentrifugation. *Instrumental Methods of Experimental.Biology*, edited by NEWMAN D. W., pp. 211–297.

TRAUTMAN R. (1966) The impact of computers on ultracentrifugation. *Fractions*, No. 2. Spinco Division of Beckman Instruments, Inc., Palo Alto, California.

VAN HOLDE K. E. and BALDWIN R. L. (1958) Rapid attainment of sedimentation equilibrium. *J. Phys. Chem.* **62**, 734–743.

VINOGRAD J., GREENWALD R. and HEARST J. F. (1965) Effect of temperature on the buoyant density of bacterial and viral DNA in CsCl solutions in the ultracentrifuge. *Biopolymers*, **3**, 109–114.

VINOGRAD J. and HEARST J. E. (1962) Equilibrium sedimentation of macromolecules and viruses in a density gradient. *Prog. in the Chem. of Organic Natural Products* **20**, 372–422.

WILLIAMS A. and IFFT J. B. Buoyant density titration of bovine serum mercaptalbumin in CsCl solutions (in preparation).

WILLIAMS J. W., VAN HOLDE K. E., BALDWIN R. L. and FUJITA H. (1958) The theory of sedimentation analysis. *Chem. Rev.* **58**, 715–806.

WRIGHT R. R., PAPPAS W. S., CARTER J. A. and WEBER C. W. (1966) Preparation and recovery of cesium compounds for density gradient solutions. *The Development of Zonal Centrifuges*, edited by ANDERSON N. G. National Cancer Institute Monograph, No. 21, pp. 241–249, U.S. Dept. Health, Education and Welfare, Bethesda.

YEANDLE S. (1958) Effect of electric field on equilibrium sedimentation of macromolecules in a density gradient of cesium chloride. *Proc. Natl. Acad. Sci. U.S.* **45**, 184–188.

ZILIUS J. and IFFT J. B. The buoyant densities of several proteins in a variety of solvents (unpublished).

AUTHOR INDEX

SUBJECT INDEX